This report contains the collective views of an international group of experts and does not necessarily represent the decisions or the stated policy of the United Nations Environment Programme, the International Labour Organisation, or the World Health Organization

Environmental Health Criteria 69

MAGNETIC FIELDS

Published under the joint sponsorship of
the United Nations Environment Programme,
the International Labour Organisation,
and the World Health Organization

World Health Organization
Geneva, 1987

The **International Programme on Chemical Safety (IPCS)** is a joint venture of the United Nations Environment Programme, the International Labour Organisation, and the World Health Organization. The main objective of the IPCS is to carry out and disseminate evaluations of the effects of chemicals on human health and the quality of the environment. Supporting activities include the development of epidemiological, experimental laboratory, and risk-assessment methods that could produce internationally comparable results, and the development of manpower in the field of toxicology. Other activities carried out by the IPCS include the development of know-how for coping with chemical accidents, coordination of laboratory testing and epidemiological studies, and promotion of research on the mechanisms of the biological action of chemicals.

ISBN 92 4 154269 1

ISSN 0250-863X

PRINTED IN FINLAND

87/7307 — VAMMALA — 6000

CONTENTS

Page

Page

WHO/IRPA TASK GROUP ON MAGNETIC FIELDS

Members

Dr V. Akimenko, A.N. Marzeev Research Institute of General and Communal Hygiene, Kiev, USSR
Dr B. G. Bernardo, Philippine Atomic Energy Commission, Quezon City, Philippines
Professor J. Bernhardt, Institute for Radiation Hygiene of the Federal Health Office, Neuherberg, Federal Republic of Germany[a]
Dr B. Bosnjakovic, Ministry of Housing, Planning and Environment, Directorate of Radiation Protection, Stralenbescherming Leidschendam, Netherlands[a]
Mrs A. Duchêne, Commissariat à l'Energie Atomique, Département de Protection Sanitaire, Fontenay-aux-Roses, France[a]
Professor J. Dumansky, A. N. Marzeev Research Institute of General and Communal Hygiene, Kiev, USSR
Professor M. Grandolfo, Radiation Laboratory, Higher Institute of Health, Rome, Italy[a]
Dr H. Jammet, Commissariat à l'Energie Atomique, Commissariat à l'Energie Atomique, Institut de Protection et de Sûreté Nucléaire, Fontenay-aux-Roses, France (Co-Chairman)[a]
Dr Y. A. Kholodov, Institute of Higher Nervous Activity and Neurophysiology, Moscow, USSR
Professor B. Knave, Research Department, National Board of Occupational Safety and Health, Solna, Sweden[a]
Dr S. Mohanna, Radiation Protection Bureau, Environmental Health Directorate, Ottawa, Ontario, Canada
Dr M. H. Repacholi, Royal Adelaide Hospital, Adelaide, South Australia (Rapporteur)[a]
Dr R. D. Saunders, National Radiological Protection Board, Chilton, Didcot, United Kingdom
Professor M. G. Shandala, A.N. Marzeev Research Institute of General and Communal Hygiene, Kiev, USSR (Co-Chairman)
Mr J. Skvarca, National Direction of Environmental Quality, Ministry of Health and Social Action, Buenos Aires, Argentina
Mr D. Sliney, Laser Microwave Division, US Army Environmental Hygiene Agency, Aberdeen Proving Ground, Maryland, USA[a]
Dr T.S. Tenforde, Lawrence Berkeley Laboratory, Biology and Medicine Division, Berkeley, California, USA

---

[a] From the Committee on Non-Ionizing Radiation of the International Radiation Protection Association.

Secretariat

Dr M. Swicord, Division of Diagnostic, Therapeutic and Rehabilitative Technology, World Health Organization, Geneva, Switzerland (WHO Consultant)
Dr P. J. Waight, Prevention of Environmental Pollution, World Health Organization, Geneva, Switzerland (Secretary)

Observers

Dr Zh.I. Chernaya, A. N. Marzeev Research Institute of General and Communal Hygiene, Kiev, USSR
Dr V. Voronin, Centre of International Projects, Moscow, USSR
Dr Z. Grigorevskaya, Centre of International Projects, Moscow, USSR
Dr T. Lukina, Centre of International Projects, Moscow, USSR

NOTE TO READERS OF THE CRITERIA DOCUMENTS

Every effort has been made to present information in the criteria documents as accurately as possible without unduly delaying their publication. In the interest of all users of the environmental health criteria documents, readers are kindly requested to communicate any errors that may have occurred to the Manager of the International Programme on Chemical Safety, World Health Organization, Geneva, Switzerland, in order that they may be included in corrigenda, which will appear in subsequent volumes.

* * *

PREFACE

The International Radiation Protection Association (IRPA) initiated activities concerned with non-ionizing radiation by forming a Working Group on Non-Ionizing Radiation in 1974. This Working Group later became the International Non-Ionizing Radiation Committee (IRPA/INIRC), at the IRPA meeting held in Paris in 1977. The IRPA/INIRC reviews the scientific literature on non-ionizing radiation and makes assessments of the health risks of human exposure to such radiation. Based on the Environmental Health Criteria Documents developed in conjunction with the International Programme on Chemical Safety (IPCS), World Health Organization, the IRPA/INIRC recommends guidelines on exposure limits, drafts codes of safe practice, and works in conjunction with other international organizations to promote safety and standardization in the non-ionizing radiation field.

The first draft of this document was compiled by DR M. REPACHOLI. An editorial group chaired by DR P. CZERSKI and including DR V. AKIMENKO, PROFESSOR J. BERNHARDT, DR B. BOSNJAKOVIC, MRS A. DUCHENE, PROFESSOR M. GRANDOLFO, DR M. REPACHOLI, MR D. SLINEY, and DR T. TENFORDE met in Neuherberg in May 1985 to develop the second draft. A small editorial group consisting of DR P. CZERSKI, DR M. SWICORD, and DR P. WAIGHT met in Geneva in April 1986 to collate and incorporate the comments received from IPCS Focal Points and individual experts. The final draft was then sent to WHO/IRPA Task Group members and formally reviewed in Kiev, USSR, 30 June - 4 July 1986. Final scientific editing of the document was completed by DR M. REPACHOLI, with the assistance of DR M. SWICORD, in Geneva in July 1986. The scientific assistance and helpful comments of DR T. TENFORDE, and the permission to use his extensive literature files, are gratefully acknowledged.

This document comprises a review of data of effects of magnetic field exposure on biological systems, pertinent to the evaluation of health risks for man. The purpose of the document is to provide an overview of the known biological effects of magnetic fields, to identify gaps in this knowledge so that direction for further research can be given, and to provide information for health authorities and regulatory agencies on the possible effects of magnetic-field exposure on human health, so that guidance can be given on the assessment of risks from occupational and general population exposure.

Subjects reviewed include: the physical characteristics of magnetic fields; measurement techniques; applications of magnetic fields and sources of exposure; mechanisms of interaction; biological effects; and guidance on the development of protective measures, such as regulations or safe-use guidelines. Health agencies and regulatory authorities are encouraged to set up and develop programmes that ensure that the maximum benefit occurs with the lowest exposure. It is hoped that this criteria document will provide useful information for the development of national protection measures against magnetic fields.

The WHO Regional Office for Europe prepared a publication entitled Non-Ionizing Radiation Protection (WHO, 1982). A revised and updated edition, completed in 1986, includes a section (5) on Electrical and Magnetic Fields at Extremely Low Frequencies.

1. SUMMARY, CONCLUSIONS, AND RECOMMENDATIONS FOR FURTHER STUDIES

This document includes a detailed review and evaluation of data on effects on human beings and other biological systems exposed to static magnetic fields or to time-varying fields at extremely low frequencies (ELF) of up to about 300 Hz. Data from the biological effects of exposure to sinusoidally varying fields are mainly concerned with effects in the range up to 20 Hz or at 50 and 60 Hz, and only limited data are available on effects at higher frequencies. Data on studies with higher frequencies and pulse repetition rates, and non-sinusoidal waveforms have also been considered, but radio-frequency magnetic fields in the frequency range 100 kHz - 300 GHz have been excluded because these have been treated in the Environmental Health Criteria 16: Radiofrequency and microwaves (WHO, 1981).

Information for health authorities on the biological effects and possible health effects of magnetic fields, is given to provide guidance for the assessment of the occupational and public health significance of exposure to magnetic fields and to indicate areas that may be hazardous. Information on human exposure levels is provided, which, on the basis of present knowledge, is considered appropriate for the prevention of health hazards.

1.1 Physical Characteristics and Dosimetric Concepts

A magnetic field always exists when there is an electric current flowing. A static magnetic field is formed in the case of direct current, and a time-varying magnetic field is produced by alternating current sources.

The fundamental vector quantities describing a magnetic field are field strength, H (unit: A/m) and magnetic flux density, B (unit: T, tesla). These quantities are related through $B = \mu H$, where μ is the magnetic permeability of the medium.

The term "dosimetry" is used to quantify exposure. Present understanding of interaction mechanisms is insufficient to develop anything but preliminary dosimetric concepts for static or ELF magnetic fields.

In practical radiation protection, it is useful to consider static and time-varying magnetic fields separately. In the case of static magnetic fields, protection limits tend to be stated primarily in terms of the external field strength or magnetic flux density and the duration of exposure. Since time-varying magnetic fields induce eddy currents within the body, evaluation may be based on the electric eddy current density (electric field strength) in critical organs. Derived protection limits can then be expressed as exposures to external magnetic fields, whereby field strength, pulse shape (rise and decay time) and frequency, orientation of the body, and duration of the exposure need to be specified.

1.2 Natural Background and Man-Made Magnetic Fields

The natural magnetic field consists of a component originating in the earth, acting as a permanent magnet, and several small components with different spectral characteristics. At the surface of the earth, the vertical component of the permanent field is maximal at the magnetic poles, amounting to about 6.7×10^{-5} T (67 μT), and is zero at the magnetic equator; the horizontal component is maximal at the magnetic equator, amounting to about 3.3×10^{-5} T (33 μT), and is zero at the magnetic pole. The flux density of the natural time-varying fields decreases from about 10^{-7} to 10^{-14} T when the frequency of the atmospheric electromagnetic fields increases from about 0.1 Hz to 3 kHz.

The magnetic fields from man-made sources generally have higher intensities than the naturally occurring fields. In the home and public places, magnetic flux densities ranging from 0.03 μT to 30 μT are produced around household appliances, and up to 35 μT near transmission lines (50 and 60 Hz), depending on the current carried and the distance from the line. For magnetically-levitated transportation systems, static magnetic fields of 6 - 60 mT are expected in the region of a passenger's head. Security systems in libraries and storehouses operate at frequencies of between 0.1 and 10 kHz and produce fields of up to about 1 mT.

Occupational exposure to magnetic fields is mainly encountered in industrial processes involving high electric current equipment, in certain new technologies for energy production and storage, and in specialized research facilities. Around various types of welding machines, furnaces, and induction heaters, the magnetic flux densities at the operator location range from about 1 μT to more than 10 mT, depending on the magnetic field frequency and the distance from the coil. Compared to devices operating at high frequencies, lower frequency induction heaters expose operators to higher magnetic flux densities. At operator-accessible locations in industries using electrolytic processes, the mean static field level is about 5 - 10 mT.

In areas accessible to operations personnel in thermonuclear magnetic fusion and magnetohydrodynamic generating systems, the static magnetic field flux densities may reach 50 mT. Similar field strengths occur near special research facilities, e.g., bubble chambers. Typical values for the magnetic flux density at work-places near 50 or 60 Hz overhead transmission lines, substations, and in power stations are up to 0.05 mT.

In medical practice, exposure to magnetic fields results mainly from the use of magnetic resonance (MR) imaging or spectroscopy methods for diagnostic purposes or from devices generating magnetic fields for therapeutic purposes. In the MR-devices in use at present, the patient is exposed to stationary magnetic fields with intensities of up to 2 T and, during examinations, to time-varying magnetic fields as high as 20 T/s. However, most patients are not exposed to time-varying fields exceeding 1.5 T/s. The peak exposure value for the patient caused by therapeutic magnetic devices is of the order of 0.1 - 2.5 mT.

The increasing use of magnetic field-producing equipment in industrial processes, research facilities, energy production and distribution, new transportation technologies, consumer products and medical practice, increases the possibility of human exposure to magnetic fields. Although, up to now, both occupational and general-population exposures to magnetic fields have generally been at low levels, some new technologies, e.g., magnetically-levitated trains, might result in exposure of the general population to levels comparable with the highest ones in some working environments. Thus, new technologies involving the production of magnetic fields should be carefully evaluated with respect to potential health risks.

1.3 Field Measurement

In order to adequately characterize a magnetic field, the magnitude, frequency, and direction of the field must be determined. The spatial properties of the field can become complicated by time-varying changes in the direction of the resultant magnetic field vector. For example, for a circularly polarized field, the magnetic vector describes an ellipse during the course of a cycle and does not reach zero magnitude. Principles of calculation and measurement of these fields are outlined.

A human or animal body located in a magnetic field causes virtually no perturbation of the field. A time-varying magnetic field induces electric currents in the exposed body. The factors affecting the magnitude of the induced currents are discussed below.

1.4 Biological Interactions

The following topics are summarized: the present state of knowledge on the mechanisms by which magnetic fields interact with living systems, and the biological effects of these fields. On the basis of available information, the areas of future research that appear to hold the greatest potential for elucidating some poorly understood aspects of magnetic field interactions with biological systems are given at the end of this section.

1.4.1 Interaction mechanisms

There are three established physical mechanisms through which static and ELF magnetic fields interact with living matter:

A. Magnetic induction

This mechanism is relevant to both static and time-varying fields, and originates through the following types of interaction:

(a) Electrodynamic interactions with moving electrolytes

Both static and time-varying fields exert Lorentz forces on moving ionic charge carriers, and thereby give rise to induced electric fields and currents. This interaction is the basis of magnetically-induced blood flow potentials that have been studied with both static and time-varying ELF fields. It is also the physical basis of the weak induced potentials that provide sensory directional cues to elasmobranch fish as they swim through the static geomagnetic field.

(b) Faraday currents

Time-varying magnetic fields induce currents in living tissues in accordance with the Faraday law of induction. Available evidence suggests that this mechanism may underlie the visuosensory stimulation that produces magnetophosphenes and other effects on electrically excitable tissues. In addition, indirect evidence suggests that rapidly time-varying magnetic fields may exert effects on a variety of cellular and tissue systems by inducing local currents that exceed the naturally occurring levels. This effect may be the basis for the wide spectrum of biological perturbations that have been observed with pulsed magnetic fields, such as those used clinically for bone fracture reunion.

B. Magnetomechanical effects

The two types of mechanical effects that a static magnetic field exerts on biological objects are:

(a) Magneto-orientation

In a uniform static field, both diamagnetic and paramagnetic molecules experience a torque, which tends to orientate them in a configuration that minimizes their free energy within the field. This effect has been well studied for assemblies of diamagnetic macromolecules with differing magnetic susceptibilities along the principal axes of symmetry. Included in this class of macromolecules are the arrays of photopigments in retinal rod disc membranes.

(b) Magnetomechanical translation

Spatial gradients of static magnetic fields produce a net force on paramagnetic and ferromagnetic materials that leads to translational motion. Because of the limited amount of magnetic material in most living objects, the influence of this effect on biological functions is negligible.

C. Electronic interactions

Certain classes of chemical reactions involve radical electron intermediate states in which interactions with a static magnetic field produce an effect on electronic spin states. It is possible, that the usual lifetime of biologically relevant electron intermediate states is sufficiently short that magnetic field interactions exert only a small, and perhaps negligible, influence on the yield of chemical reaction products.

In addition to the mechanisms of magnetic field interactions for which there is direct experimental evidence, several other mechanisms have been proposed, on theoretical grounds, in an effort to explain various biological effects that have been reported to occur in static and ELF fields of very low intensity. However, it must be emphasized, that many proposed mechanisms have not been subjected to direct experimental tests.

1.4.2 Biological effects of magnetic fields

Some organisms possess sensitivity to static magnetic fields with low intensities comparable to that of the geomagnetic field (about 50 μT). Phenomena for which there is substantial experimental evidence of sensitivity to the earth's field include:

(a) direction finding by elasmobranch fish (shark, skate, and ray);

(b) orientation and swimming direction of magnetotactic bacteria;

(c) kinetic movements of molluscs;

(d) migratory patterns of birds; and

(e) waggle dance of bees.

In addition, a number of in vitro studies have been made of magnetic orientation in assemblies of macromolecules, including retinal rod outer segments, muscle fibres, photosynthetic systems (chloroplast grana, photosynthetic bacteria, and Chlorella cells), halobacteria purple membranes, and various synthetic liquid crystals and gels. As discussed in the preceding summary of mechanisms of magnetic field interaction, certain classes of chemical reactions that involve a radical electron intermediate state may also be sensitive to static magnetic fields of moderate intensity (< 10 mT).

The available experimental information on the response of organisms, including land-dwelling mammalian species, to static and ELF magnetic fields indicates that three biological effects can be regarded as established phenomena:

(a) the induction of electrical potentials within the circulatory system;

(b) magnetophosphene induction by pulsed and ELF magnetic fields with a time rate of change exceeding 1.3 T/s or sinusoidal fields of 15 - 60 Hz and field strengths ranging from 2 to 10 mT (frequency dependent); and

(c) the induction by time-varying fields of a wide variety of cellular and tissue alterations, when the induced current density exceeds 10 mA/m^2; many of these effects appear to be the consequence of interactions with cell membrane components.

For static magnetic fields with flux densities of less than 2 T, there exists a body of experimental data that indicates the absence of irreversible effects on many developmental, behavioural, and physiological parameters in higher organisms. Broadly summarized, available evidence suggests that the following 9 classes of biological functions are not significantly affected by static magnetic fields at levels up to 2 T:

(a) cell growth;

(b) reproduction;

(c) pre- and post-natal development;

(d) bioelectric activity of isolated neurons;

(e) behaviour;

(f) cardiovascular functions (acute exposures);

(g) the blood-forming system and blood;

(h) immune system functions;

(i) physiological regulation and circadian rhythms.

For time-varying magnetic fields in the ELF frequency range, few systematic studies have been carried out to define the threshold field characteristics for producing significant perturbations of biological functions. Nevertheless, available evidence suggests that ELF magnetic fields must induce current densities in tissues and extracellular fluids that exceed 10 mA/m², in order to produce significant alterations in the development, physiology, and behaviour of intact higher organisms. In <u>in vitro</u> studies, various phenomena have been reported in the 1 - 10 mA/m² range, but their health significance has not been determined. However, it should be noted that therapeutic applications of magnetic fields make use of this range.

1.5 Effects on Man

1.5.1 Static fields

Studies on workers involved in the manufacture of permanent magnets in the USSR indicated various subjective symptoms and functional disturbances including irritability, fatigue, headache, loss of appetite, bradycardia, tachycardia, decreased blood pressure, altered EEG, itching, burning, and numbness. However, lack of any statistical analysis or assessment of the impact of physical or chemical hazards in the working environment significantly reduces the value of these reports. Although the studies are inconclusive, they suggest that, if long-term effects occur, they are very subtle, since no cumulative gross effects are evident.

Recent epidemiological surveys in the USA have failed to reveal any significant health effects associated with long-term exposure to static magnetic fields. A study of the health data on 320 workers in plants using large electrolytic cells for chemical separation processes, where the average static field level in the work environment was 7.6 mT and the maximum field was 14.6 mT, indicated slight changes in white blood cell picture (still within the normal range) in the exposed group compared with the 186 controls. None of the observed changes in blood pressure or blood parameters was considered indicative of a significant adverse effect associated with magnetic field exposure.

The prevalence of disease among 792 workers at the US National Accelerator Laboratories, who were exposed occupationally to static magnetic fields, was compared with that in a control group consisting of 792 unexposed workers matched for age, race, and socioeconomic status. The range of magnetic field exposures was from 0.5 mT for long durations to 2 T for periods of several hours. No significant increase or decrease in the prevalence of 19 categories of disease was observed in the exposed group relative to the controls.

Workers exposed to large static magnetic fields in the aluminium industry were reported to have an elevated leukaemia mortality rate. Although these studies suggest an increased cancer risk for persons directly involved in aluminium production, there is no clear evidence, at present, indicating the responsible carcinogenic factors within the work environment.

It can be concluded that available knowledge indicates the absence of any adverse effects on human health due to exposure to static magnetic fields up to 2 T. It is not possible to make any definitive statements about safety or hazard associated with exposure to fields above 2 T. From theoretical considerations and some experimental data, it could be inferred that short-term exposure to static fields above 5 T may produce significant detrimental effects on health.

1.5.2 Time-varying fields

Time-varying magnetic fields generate internal electric currents. For example, 3 T/s can induce current densities of about 30 $\mu A/m^2$ around the perimeter of the human head. Induced electric current densities can be used as the decisive parameter in the assessment of the biological effects at the cellular level.

In terms of a health risk assessment, it is difficult to correlate the internal tissue current densities with the external magnetic field strength. However, assuming worst-case conditions, it is possible to calculate, at least within one order of magnitude, the magnetic flux density that would produce potentially hazardous current densities in tissues. The following statements can be made on induced current density ranges and correlated magnetic flux densities of a sinusoidal homogeneous field, which produce biological effects from whole-body exposure:

(a) Between 1 and 10 mA/m² (induced by magnetic fields above 0.5 - 5 mT at 50/60 Hz, or 10 - 100 mT at 3 Hz), minor biological effects have been reported.

(b) Between 10 and 100 mA/m² (above 5 - 50 mT at 50/60 Hz or 100 - 1000 mT at 3 Hz), there are well established effects, including visual and nervous system effects. Facilitation of bone fracture reunion has been reported.

(c) Between 100 and 1000 mA/m² (above 50 - 500 mT at 50/60 Hz or 1 - 10 T at 3 Hz), stimulation of excitable tissue is observed and there are possible health hazards.

(d) above 1000 mA/m² (greater than 500 mT at 50/60 Hz or 10 T at 3 Hz), extra systoles and ventricular fibrillation, i.e., acute health hazards, have been established.

For non-sinusoidal waveforms that have short duration pulses, the time rate of change of the magnetic flux density must be specified. In analysing certain biological effects, especially the stimulation of excitable tissue, the peak current density values are more relevant than root mean square (rms) values. In addition, non-homogeneous magnetic fields must be considered, since high field gradients exist near strong magnetic field sources. The induction loops in extremities are usually smaller than those in the whole body, so higher magnetic field strengths are tolerable for extremities than for the whole body.

Several laboratory studies have been conducted with human subjects exposed to sinusoidally time-varying magnetic fields with frequencies in the ELF range. None of these investigations has revealed adverse clinical or psychological changes in the exposed subjects. The strongest field used in these studies with human volunteers was a 5-mT, 50-Hz field to which subjects were exposed for 4 h.

Several recent epidemiological reports present preliminary data indicative of an increase in the incidence of cancer among children, adults, and occupational groups. In other epidemiological studies in the USA, no apparent increases in genetic defects or abnormal pregnancies were reported. The studies that show an excess of cancers in children and adults suggest an association with exposure to very weak (10^{-7} - 10^{-6} T) 50 or 60 Hz magnetic fields that are of a magnitude commonly found in the environment. These associations cannot be satisfactorily explained by the available theoretical basis for carcinogenesis by ELF electromagnetic fields. The preliminary nature of the epidemiological evidence, and the relatively small increment in reported incidence, suggest that, although these epidemiological data cannot be dismissed, there must be considerable further study before they can be accepted.

From the available data on human exposure to time-varying magnetic fields, it can be concluded that induced current densities below 10 mA/m^2 have not been shown to produce any significant biological effects. In the range of 10 - 100 mA/m^2 (from fields higher than 5 - 50 mT at 50/60 Hz), biological effects have been established, but these induced current densities from short-term exposure (few hours) may cause minor transient effects on health. The health consequences of exposure to these levels for many hours, days, or weeks are not known at present. Above 100 mA/m^2 (greater than 50 mT at 50/60 Hz), various stimulation thresholds are exceeded and hazards to health may occur.

1.6 Exposure Guidelines and Standards

Standards or guidelines limiting human exposure to static ELF magnetic fields have been developed in a few countries. Of particular interest is the increasing tendency of countries to limit magnetic field exposure from particular devices (e.g., magnetic resonance diagnostic techniques). Details of these standards and guidelines are given in section 9 of the document.

1.7 Protective Measures

Two aspects of magnetic field safety that deserve special attention are the potential influence of these fields on the functioning of electronic devices, and the risk of injury due to the large forces exerted on ferromagnetic objects in strong static magnetic field gradients. Of particular concern is the malfunction of cardiac pacemakers and the displacement of aneurysm clips and prosthetic devices.

1.7.1 Cardiac pacemakers

Both static and time-varying magnetic fields can interfere with the proper functioning of modern demand pacemakers. Some pacemakers may revert from a synchronous to an asynchronous mode of operation in time-varying fields with time rates of change above approximately 40 mT/s. Certain pacemaker models also exhibit abnormal operation due to closure of a reed relay switch in static magnetic fields that exceed 1.7 - 4.7 mT. Magnetic fields can also affect the functioning of other medical electronic monitoring devices, such as EEG and ECG equipment.

1.7.2 Metallic implants

The sensitivity of implanted surgical devices to magnetic fields is dependent on their alloy composition. A large number of metallic devices such as intrauterine devices, surgical clips, prostheses, infusion needles, and catheters may have a significant torque exerted on them by intense magnetic field gradients. This may result in their displacement and produce serious consequences. All persons entering magnetic field environments should be screened carefully and, if necessary, prohibited from access.

1.7.3 Hazards from loose paramagnetic objects

Depending on the weight and shape of a paramagnetic object subject to an intense magnetic field, it can become a missile with high momentum. Care should be taken to exclude such objects as, for example, scissors, scalpels, and handtools from the vicinity of strong magnetic field sources.

1.8 Recommendations for Future Research

On the basis of present knowledge of magnetic field bioeffects, several key areas of future research can be identified as being essential for achieving a comprehensive understanding of the biological consequences of exposure to these fields. No attempt has been made to list all possible research areas. Instead, emphasis has been placed on areas considered to have an impact on health hazard assessment.

For static magnetic fields, there is a clear need for additional studies in the following areas, in each of which the available information is either inadequate or contradictory:

(a) studies on functional alterations in the cardiovascular and central nervous system, where magnetic field interactions have previously been observed; particular emphasis should be placed on the effects of long-term exposures;

(b) sensitivity of enzyme reactions that involve radical intermediate states, which may be an important issue in long-term occupational exposures;

(c) cellular, tissue, and animal responses to static fields above 2 T, as proposed for use in clinical MR spectroscopy.

For time-varying magnetic fields with repetition frequencies in the ELF range, key areas of future research can also be recommended on the basis of available information:

(a) Comprehensive epidemiological studies should be carried out to resolve the issue of whether an elevated risk of leukaemia and other forms of cancer is associated with occupational and residential exposure to ELF fields. These studies should include the use of appropriate techniques for the assessment of field exposure parameters (e.g., the use of miniature personal dosimeters). Relevant research with cellular and animal systems should also be conducted in an effort to elucidate interaction mechanisms of ELF fields that could lead to an elevated cancer risk.

(b) Studies on the response of developing embryonic and fetal systems, and other cell and tissue systems that have been identified as being responsive to ELF magnetic fields, should be continued with particular focus on effects mediated via interactions with cell membranes.

(c) Studies are needed on the effects of low levels of induced current density (< 100 mA/m^2) on nerve tissue.

2. PHYSICAL CHARACTERISTICS, DOSIMETRIC CONCEPTS, AND MEASUREMENT

Just as an electric field is always linked with an electric charge, a magnetic field always appears when electric current flows. A magnetic field can be illustrated by lines of force. A static magnetic field is formed in the case of direct current, whereas a time-varying magnetic field is induced by alternating current sources.

The electric (E) and magnetic (H) fields that exist near sources of electromagnetic fields must be considered separately, because the very long wavelength (thousands of kilometres) characteristic of extremely low frequencies (ELF) means that measurements are made in the non-radiating near field. The E and H fields do not have the same constant relationship that exists in the far field of a radiating source.

A description of the physical characteristics of static and ELF magnetic fields has been given by Grandolfo & Vecchia (1985a). An animal or human body does not appreciably distort a magnetic field. Time-varying magnetic fields induce currents within the body. The magnitude of these internal currents is determined by the radius of the current path, the frequency of the magnetic field and its intensity at the location within the body. Unlike the electric field for which the internal field strength is many orders of magnitude less than that of the external field, the magnetic field strength is virtually the same outside the body as within. The magnetically-induced electric field strengths and corresponding current density are greatest at the periphery of the body where the conducting paths are longest, whereas microscopic current loops anywhere within the body would have extremely small current densities. The magnitude of the current density is also influenced by tissue conductivity where the exact paths of the current flow depend in a complicated way on the conducting properties of the various tissues.

2.1 Quantities and Units

The quantities, units, and symbols used in describing magnetic fields are given in Table 1.

The fundamental vector quantities describing a magnetic field are the field strength (**H**) and the magnetic flux density (**B**) (or equivalently, the magnetic induction).

The magnetic field strength (**H**) is the force with which the field acts on an element of current situated at a particular point. The value of **H** is measured in ampere per metre (A/m). The trajectories of the motion of an element of

Table 1. Magnetic field quantities and units in the SI System

Quantity	Symbol	Unit
Frequency	f	hertz (Hz)
Current	I	ampere (A)
Current density	**J**	ampere per square metre (A/m^2)
Magnetic field strength	**H**	ampere per metre (A/m)
Magnetic flux	Φ	weber (Wb) = Vs
Magnetic flux density	**B**	tesla (T) = Wb/m^2
Permeability	μ	henry per metre (H/m)
Permeability of vacuum	μ_o	$\mu_o = 1.257 \times 10^{-6}$ H/m
Time	t	seconds (s)

current (or the orientations of an elementary magnet) in a magnetic field are called the magnetic lines of force.

As in the case of electric fields, single-phase and three-phase magnetic fields can be defined: the field at any point may be described in terms of its time-varying magnitude and invariant direction (single-phase), or by the field ellipse, i.e., the magnitude and direction of the major and minor semi-axes (three phase).

The magnetic flux density (**B**), rather than the magnetic field strength, (**H** = **B**/μ), is used to describe the magnetic field generated by currents in the conductors of transmission lines and substations. Thus, the magnetic field is defined as a vector field of magnetic flux density **B** (B-field). The value of μ (the magnetic permeability) is determined by the properties of the medium, and, for most biological material is equal to μ_o, the value of the permeability of free space (air). Thus, for biological materials the values of **B** and **H** are related by a constant (μ_o).

Before the introduction of the International System of units (SI), the use of the CGS system (based on the three independent quantities: length (cm), mass (g) and time (s)) was customary. SI is based on seven independent quantities: length (m), mass (g), time (s), electric current (A), thermodynamic temperature (K), luminous intensity (cd), and amount of substance (mol). The equations describing the electromagnetic phenomena are equivalent but not identical in the SI and the CGS systems. For an electromagnetic field, only the first four of the seven quantities mentioned above, are relevant. The CGS unit of magnetic field strength is the oersted and that of the magnetic induction is the gauss.

In the CGS system, μ_0 is a dimensionless quantity equal to unity, and as a result, for biological materials, **B** can be set equal to **H**, as a close approximation. This convention has been used extensively in the biological literature, where many authors have used **B** and **H** as interchangeable quantities. Thus, many publications contain equations that are appropriate for use only with the CGS system of units since the permeability of free space, μ_0, has been omitted.

The SI system has now been universally accepted. The CGS system is obsolete and should not be used.

In addition, the term gamma is used and is equal to 1 nanotesla (10^{-9} tesla). For convenience, the conversion factors relating the various quantities used in laboratory practice are given in Table 2.

Table 2. Conversion factors for units

To obtain / To convert	T = Wb/m²	G	γ	A/m	Oe
T = Wb/m²	1	10^4	10^9	7.96×10^5	10^4
G	10^{-4}	1	10^5	79.6	1
γ	10^{-9}	10^{-5}	1	7.96×10^{-4}	10^{-5}
A/m	1.256×10^{-6}	1.256×10^{-2}	1256	1	1.256×10^{-2}
Oe	10^{-4}	1	10^5	79.6	1

Symbols: T = tesla
Wb = weber
G = gauss
A = ampere
m = metre
γ = gamma
Oe = oersted

For a more complete inventory and discussion of quantities and units, the reader is referred to a report of the IRPA/International Non-Ionizing Radiation Committee entitled "Review of Concepts, quantities, units, and terminology for non-ionizing radiation protection" (IRPA, 1985).

2.2 Dosimetric Concepts

In its broadest sense, the term "dosimetry" is used to quantify exposure to radiation. Quantitative descriptions of exposure, for the purpose of formulating protection standards and exposure limits, require the use of appropriate quantities. "Appropriate" means that the quantities should represent, as far as possible, the physical processes that are closely linked to the biological effects of the fields. Since our knowledge of interaction mechanisms is incomplete, exposure conditions are often quantified in terms of the unperturbed external magnetic field strength and the duration of exposure.

The known physical mechanisms by which magnetic fields interact with living matter are described in section 4. Some factors affecting the interaction of fields with organisms are summarized in Table 3. To fully assess the data obtained in bioeffects research, exposure conditions must be well controlled and measured. In this case, the "dosimetry" in bioeffects research with magnetic fields is very complex, since all relevant factors must be taken into account. The accuracy and sophistication of radiation protection dosimetry must be related to the conditions and actual or potential adverse consequences of exposure to magnetic fields.

Table 3. Factors affecting interaction of magnetic fields

Parameters of the magnetic field source

1. Frequency
2. Modulation (Pulse, AM, FM), rise and decay times (dB/dt)
3. Polarisation
4. Field strength
5. Field pattern (uniformity)
6. Surrounding material properties

Parameters related to exposure

1. Tissue properties (conductivity, anisotropy, permeability)
2. Size, geometry
3. Orientation relative to polarization
4. Mode of exposure (partial; whole body)

Extraneous factors

1. Metal implants (ferromagnetic)
2. Metal objects in the field
3. Drugs (medications)
4. Chemical pollutants

In practical radiation protection, it is useful to consider static and time-varying magnetic fields separately.

2.2.1 Static magnetic fields

In the assessment of exposure to static magnetic fields for practical radiation protection purposes, the appropriate quantities are less well defined. Protection limits tend to be stated in terms of the external field strength and the duration of exposure, where the integrated product of field and exposure time could be considered as a measure of exposure. However, at present, there is no biological basis for choosing this dosimetric concept. Further development of dosimetric concepts and their theoretical and experimental basis is required.

2.2.2 Time-varying magnetic fields

In evaluating human exposure to time-varying magnetic fields of frequencies between about 10 Hz and 100 kHz, the electric eddy current density can be employed as the decisive parameter in assessment of the biological effects at the cellular level (Bernhardt, 1979, 1985, 1986; Czerski, 1986; Tenforde, 1986a). Field strength and eddy current density are related by the specific conductivity of the medium.

By comparing the current densities, it may be possible to predict effects in human beings from those found in studies on animal and isolated cells. In this context, it is irrelevant whether the current density surrounding a cell is introduced into the body through electrodes or induced in the body by external magnetic fields. However, the current paths within the body may be different.

The evaluation of human exposure using current densities is based primarily on a concept of "dose" to the critical organs. Although this assumption is based on the most likely hypothesis, this mechanism of energy absorption in tissues should not be considered to the exclusion of all others. The parameters of internal field strength and duration should also be taken into account. Basic protection limits can be expressed in permissible current densities; derived protection limits can be expressed as exposures to external magnetic fields, where field strength, frequency, orientation of the body, and duration of exposure need to be specified. Refinements may include field gradient values, partial body exposure, etc. Induced eddy currents in organs cannot be measured, at present, under any practical conditions. Therefore, the only protection quantities that can be used to assess exposure to time-varying magnetic fields are the field strength distribution in time and space.

2.3 Measurement of Magnetic Fields

During the last thirty years, the measurement of magnetic fields has undergone considerable development. Progress in techniques has made it possible to develop new methods of measurement as well as to improve old ones. Some of the incentive for considerable development in magnetic measurement techniques has arisen because of the necessity to accurately measure magnetic fields that often vary in both space and time in large particle accelerators. The rapid development of plasma physics as well as that of astronautics has created new demands for magnetic field measurements.

A description of the most common measuring techniques follows, together with a comparison of their advantages and limitations. Further details can be found in Williamson & Kaufman (1981), Grandolfo & Vecchia (1985a), and Stuchly (1986).

The two most popular types of magnetic field probes are a shielded coil and a Hall-probe. Most of the commercially available magnetic field meters use one of them. Recently, in addition to Hall probes, other semiconductor devices, namely bipolar transistors and FET transistors, have been proposed as magnetic field sensors. They offer some advantages over Hall probes, such as higher sensitivity, greater spatial resolution, and broader frequency response.

For measurements of very weak magnetic fields, such as those produced by endogeneous currents in biological systems, other sensors are used. These include fluxgates, optically pumped magnetometers, magnetostrictive sensors with optical fibres, and superconducting quantum interference devices (SQUIDS). These devices are rather specialized and expensive and are not normally used for the measurement of extraneous fields in biomedical applications (Stuchly, 1986).

2.3.1 Search coils

The operating principle of a coil B-field probe can be explained by considering a closed loop of a conductor with area A immersed in a quasi-static, uniform magnetic field of flux density B, and angular frequency $\omega(= 2\pi f)$, as shown in Fig. 1 (Conti, 1985).

An electromotive force (EMF) is induced in the loop (and a current (I) will flow) as a consequence of changes in the magnetic flux $\Phi(B)$ through the area A, in accordance with the following law:

$$\mathrm{EMF} = - \frac{d}{dt} \Phi (B) \quad (1)$$

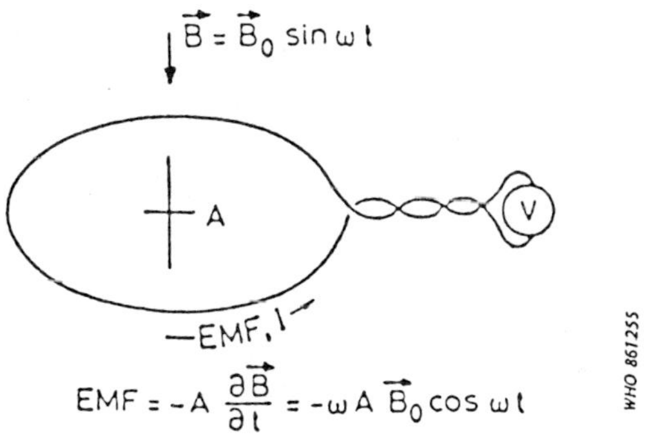

Fig. 1. Conducting loop in power frequency B-field.

If the vector **B** = B_0 sin ωt is assumed to be uniform and to have its direction perpendicular to the plane of the loop, the EMF is given by the following relationship:

$$EMF = - \frac{d}{dt} (A\ B_0 \sin\ \omega t) = \omega B_0\ A \cos\ \omega t \qquad (2)$$

Equation (2) shows that measurement of induced electromotive force provides a measure of the B-field strength.

For a loop of many turns, the EMF given by Equation (2) will develop over each turn and the voltage (V) will increase accordingly. The induced current has been assumed to be so small that the opposing B-field generated by I can be ignored.

There is no theoretical limit on the frequency of operation of coils as sensors, except for the loop size. In practice, factors such as the electric field perturbation and the pick up by the leads connecting the loop to the metering device require modifications of the sensor design.

A single coil has a directional spatial response characteristic, and has to be rotated to obtain a maximum reading to determine the actual magnitude and direction of the field. Alternatively, a probe consisting of three mutually perpendicular coils can be designed.

2.3.2 The Hall probe

The most commonly used method in field mapping is the Hall probe. When a strip of conducting material is placed along the O_x axis in a coordinate system O_{xyz}, with a current I running in the direction O_x while a magnetic field B is applied in the direction O_y at right angles to the surface of the strip, a potential difference appears in the direction O_z between the two sides of the strip.

The Hall effect can be explained as the result of the action exerted on the charge carriers by the magnetic field, which forces them sideways in the strip. Thus, electric charges appear on the sides of the strip and, as a result, a transverse Hall electric field is created.

Several factors set limits on the accuracy obtainable, the most serious being the temperature coefficient of the Hall voltage. Another complication can be that of the planar Hall effect, which makes the measurement of a weak field component normal to the plane of the Hall plate problematical, when a strong field component is present parallel to this plane. Many possible remedies have been proposed, but they are all relatively difficult to apply. Last, but not least, is the problem of the representation of the calibration curve since the Hall coefficient varies with the magnetic field.

The measurement of the Hall voltage sets a limit of about 0.1 mT on the sensitivity and resolution of the measurement, if conventional direct current excitation is applied to the probe. The sensitivity can be improved considerably by using alternating current excitation. Higher accuracy at low field strengths can be achieved by using synchronous detection techniques for the measurement of the Hall voltage.

Hall plates are usually calibrated in a magnet in which the field is measured simultaneously using a nuclear magnetic resonance probe. A well designed Hall-probe assembly can be calibrated to an accuracy of 0.01% (Germain, 1963).

2.3.3 Nuclear magnetic resonance probe

Nuclear magnetic resonance (NMR) is the classical method of measuring the absolute value of a magnetic field.

If a charged particle possessing an angular momentum vector, **J**, is placed in a constant magnetic field **B**, the magnetic moment, **u**, of the particle becomes orientated with respect to **B**. The vectors **J** and **u** are proportional, $\mathbf{u} = \gamma \mathbf{J}$ where γ is the gyromagnetic ratio of the particle considered. In a quantum mechanics description, this orientation can only be such that the component of **J** along **B** is equal to $mh/2\pi$, where $m = \pm(I - k)$, I is the spin of the particle, and k is an integer smaller or equal to I. Thus, m can take on several discrete values, each giving a different orientation for **J** and **u**. Each of these orientations of **u** in the magnetic field corresponds with a different energy level, where these levels differ in energy by $\Delta E = \mathbf{B}\ \gamma h/2\pi$.

If a sample containing a large number of particles, either electrons or protons, is irradiated with photons of the right frequency, υ_0, such that $h\upsilon_0 = \Delta E$, an exchange of energy occurs. As a result of photon absorption, particles in the sample jump from the lower to the higher energy level. The principle of the NMR measurement technique is to determine the resonant frequency of the test specimen in the magnetic field to be measured. It is an absolute measurement that can be made with very great accuracy. The measuring range of this method is from about 10^{-2} to 10 T, without definite limits.

In field measurements using the proton magnetic resonance method, an accuracy of 10^{-4} is easily obtained with simple apparatus and an accuracy of 10^{-6} can be reached with extensive precautions and refined equipment.

The inherent shortcoming of the NMR method is its limitation to fields with a low gradient and the lack of information about the field direction.

2.3.4 Personal dosimeters

A personal dosimeter suitable for monitoring exposures to static and time-varying magnetic fields has been developed by Fujita & Tenforde (1982). Using thin-film Hall sensors that record magnetic induction (B) along three orthogonal axes, the time rate of change of the magnetic induction (dB/dt) is determined for values of B recorded during consecutive sampling intervals. The parameters stored by the dosimeter include the average and peak values of B and dB/dt during a preset time interval, and the number of times that specified threshold levels of these parameters are exceeded. An audible alarm sounds when B or dB/dt exceeds a preset threshold level. This personal dosimeter is battery operated, and is capable of recording magnetic field exposure throughout an 8-h working day. A microprocessor-controlled field dosimeter for monitoring personal exposures to power-frequency magnetic fields has been developed by Lo et al. (1986). This dosimeter uses electrically-shielded, 500-turn copper coils and synchronous detector circuits for field measurements along three orthogonal axes. For 60-Hz fields a measurement accuracy of 1 - 2% is achieved over the range of magnetic flux densities from 5 nT to 60 μT (rms).

3. NATURAL BACKGROUND AND MAN-MADE MAGNETIC FIELDS

3.1 Natural Magnetic Fields

The natural magnetic field consists of one component due to the earth acting as a permanent magnet and several other small components, which differ in characteristics and are related to such influences as solar activity and atmospheric events (Aleksandrov et al., 1972; Polk, 1974; Benkova, 1975; Grandolfo & Vecchia, 1985b). The earth's magnetic field originates from electric current flow in the upper layer of the earth's core. There are significant local differences in the strength of this field. At the surface of the earth, the vertical component is maximal at the magnetic poles, amounting to about 6.7×10^{-5} T (67 μT) and is zero at the magnetic equator. The horizontal component is maximal at the magnetic equator, about 3.3×10^{-5} T (33 μT), and is zero at the magnetic pole.

The naturally occurring time-varying fields in the atmosphere have several origins, including diurnally varying fields of the order of 3×10^{-8} T (0.03 μT) associated with solar and lunar influences on ionospheric currents. The largest time-varying atmospheric magnetic fields arise intermittently from intense solar activity and thunderstorms, and reach intensities of the order of 5×10^{-7} T (0.5 μT) during large magnetic storms.

About 2000 thunderstorms are occurring simultaneously over the globe with lightning striking the earth's surface about 16 times per second; the currents involved may reach 2×10^5 A at the level of the earth (Kleimenova, 1963). Electromagnetic fields having a very broad frequency range (from a few Hz up to a few MHz) originate the moment lightning strikes and propagate over long distances influencing the magnitude of magnetic fields. Superimposed on the magnetic fields associated with irregular atmospheric events is a weak time-varying field resulting from the Schumann resonance phenomenon. These fields are generated by lightning discharges and propagate in the resonant atmospheric cavity formed by the earth's surface and the lower boundary of the ionosphere.

The characteristics of the time-varying components of the natural magnetic field can be summarized as follows:

(a) The magnetic flux densities from 5 to 10×10^{-8} T are at pulsation frequencies from 0.002 to 0.1 Hz.

(b) The geomagnetic pulsations up to 5 Hz are of short duration, lasting from a few minutes to a few hours.

(c) The magnetic flux densities of the field decrease with increasing frequency from 10^{-11} T at 5 - 7 Hz to 10^{-14} T at 3 kHz.

3.2 Man-Made Sources

The static and time-varying magnetic fields originating from man-made sources generally have much higher intensities than the naturally occurring fields. This statement is particularly true for sources operating at the power frequencies of 50 or 60 Hz (e.g., home appliances), where fields occur that are many orders of magnitude greater than the natural fields at the same frequencies. Other man-made sources are to be found in research, industrial and medical procedures, and in several other technologies related to energy production and transportation that are in the developmental stage (Demetsky & Alekseev, 1981; Stuchly, 1986; Tenforde, 1986b). A list of applications of magnetic field technologies is given in Table 4.

Table 4. Magnetic field technologies[a]

<u>Energy technologies</u>
Thermonuclear fusion reactors
Magnetohydrodynamic systems
Superconducting magnet energy storage systems
Superconducting generators and transmission lines
<u>Research facilities</u>
Bubble chambers
Superconducting spectrometers
Particle accelerators
Isotope separation units
<u>Industry</u>
Aluminium production
Electrolytic processes
Production of magnets and magnetic materials
<u>Transportation</u>
Magnetically levitated vehicles
<u>Medicine</u>
Magnetic resonance
Therapeutic applications

[a] From: Tenforde (1986b).

3.2.1 Magnetic fields in the home and public premises

3.2.1.1 Household appliances

Some common electrical appliances and the typical magnetic fields near them are listed in Table 5. In a survey of magnetic fields around almost 100 different 60-Hz household appliances, levels from 0.03 μT to 30 μT were measured at a distance of 30 cm from the device (Gauger, 1984). At approximately 150 cm from the appliance producing the highest magnetic field, the level had fallen to about 0.5 μT. Background magnetic field flux densities in the homes where the fields from appliances were measured, ranged between 0.05 to 1 μT (Tell, 1983; Male et al., 1984; Stuchly, 1986).

Table 5. Magnetic flux densities at 60 Hz near various appliances in the USA[a]

Appliance	Magnetic flux density (μT) at distance z		
	z = 3 cm	z = 30 cm	z = 1 m
Can openers	1000 - 2000	3.5 - 30	0.07 - 1
Hair dryers	6 - 2000	< 0.01 - 7	< 0.01 - 0.3
Electric shavers	15 - 1500	0.08 - 9	< 0.01 - 0.3
Sabre and circular saws	250 - 1000	1 - 25	0.01 - 1
Drills	400 - 800	2 - 3.5	0.08 - 0.2
Vacuum cleaners	200 - 800	2 - 20	0.13 - 2
Mixers	60 - 700	0.6 - 10	0.02 - 0.25
Fluorescent desk lamps	40 - 400	0.5 - 2	0.02 - 0.25
Garbage disposals	80 - 250	1 - 2	0.03 - 0.1
Microwave ovens	75 - 200	4 - 8	0.25 - 0.6
Fluorescent fixtures	15 - 200	0.2 - 4	0.01 - 0.3
Electric ranges	6 - 200	0.35 - 4	0.01 - 0.1
Portable heaters	10 - 180	0.15 - 5	0.01 - 0.25
Blenders	25 - 130	0.6 - 2	0.03 - 0.12
Television	2.5 - 50	0.04 - 2	< 0.01 - 0.15
Electric ovens	1 - 50	0.15 - 0.5	0.01 - 0.04
Clothes washers	0.8 - 50	0.15 - 3	0.01 - 0.15
Irons	8 - 30	0.12 - 0.3	0.01 - 0.025
Fans and blowers	2 - 30	0.03 - 4	0.01 - 0.35
Coffee makers	1.8 - 25	0.08 - 0.15	< 0.01
Dishwashers	3.5 - 20	0.6 - 3	0.07 - 0.3
Toasters	7 - 18	0.06 - 0.7	< 0.01
Crock pots	1.5 - 8	0.08 - 0.15	< 0.01
Clothes dryers	0.3 - 8	0.08 - 0.3	0.02 - 0.06
Refrigerators	0.5 - 1.7	0.01 - 0.25	< 0.01

[a] From: Gauger (1984).

3.2.1.2 Transmission lines

The magnetic field beneath high-voltage overhead transmission lines is mainly transversed to the line axis (Fig. 2). The maximum flux density at ground level may be under the centre line or under the outer conductors, depending on the phase relationship between the conductors. Apart from the geometry of the conductor, the maximum magnetic field strength is determined only by the magnitude of the current. The maximum magnetic flux density at ground level for a double-circuit 500 kV overhead transmission line system is approximately 35 μT per kiloampere. The field at ground level beneath a 765-kV, 60-Hz power line carrying 1 kA per phase is 15 μT (Scott-Walton et al., 1979). The magnetic flux density decreases with distance from the conductor to values of the order of 1 - 10 μT at a lateral distance of about 20 - 60 m from the line, as shown in Fig. 2 (Lambdin, 1978; Zaffanella & Deno, 1978).

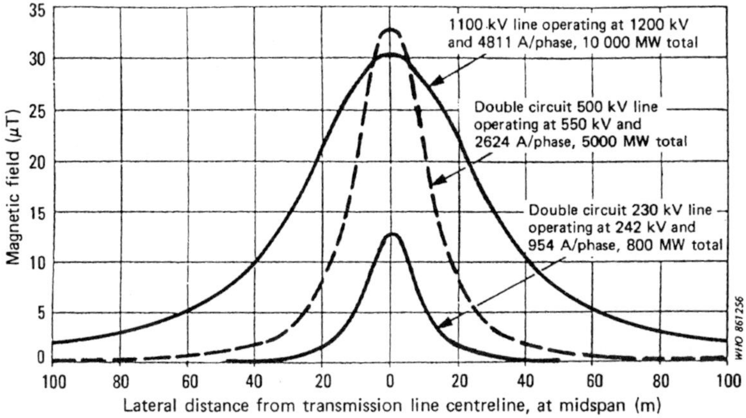

Fig. 2. Profile of calculated magnetic field strength (60 Hz) at 1 m above ground for 230-kV and 500-kV double circuit lines, and a possible design for a future 1100-kV line. From: Lee et al. (1982).

3.2.1.3 Transportation

Several countries are currently designing and testing prototype vehicles that are suspended and guided by magnetic forces. If successful, the magnetically-levitated vehicle could offer high-speed public transportation (roughly 200 - 400 km/h), with greatly reduced levels of noise and pollution, compared with conventional modes of transportation.

The technical problems of magnetic levitation are considerable and include the presence of large fringe magnetic fields within the passenger compartment. In some designs, the field level at the floor of the passenger compartment may be 50 - 100 mT, and estimates of the field at the location of a passenger's head range from 6 to 60 mT (Hassenzahl et al., 1978). The magnetic flux density within the passenger compartment can be significantly reduced by several procedures and it may be possible to achieve a 5- to 10-fold reduction in the magnetic field levels to which passengers are exposed.

3.2.1.4 Security systems

Different security systems have been developed for personnel identification or for electronic surveillance against theft in libraries and shops. Such devices operate at frequencies ranging between 0.1 and 10 kHz. Identification is achieved when a person passes through the coil carrying an identification tag or articles bearing a magnetic strip. The maximum magnetic flux generated by the coil is about 1 mT at the ground.

The maximum magnetic flux for walk-through metal detectors used at airports is below 0.1 mT, and they have frequencies of operation below 1 MHz.

3.2.2 Magnetic fields in the work-place

3.2.2.1 Industrial processes

Occupational exposure to magnetic fields comes predominantly from working near industrial equipment using high currents. Such devices include various types of welding machine, electroslag refining, various furnaces, induction heaters, and stirrers. Details of surveys of magnetic field strengths in industrial settings are given in Table 6. Surveys on induction heaters used in industry performed in Canada (Stuchly & Lecuyer, 1985), in Poland (Aniolczyk, 1981), and in Sweden (Lövsund et al., 1982), show magnetic flux densities at operator locations ranging from 0.7 μT to 6 mT, depending on the frequency used and the distance from the machine. In their study of magnetic fields from industrial electro-steel and welding equipment, Lövsund et al. (1982) found that spot welding machines (50 Hz, 15 - 106 kA) and

Table 6. Occupational sources of exposure to magnetic fields

Source	Magnetic flux densities (mT)	Distance (m)	Reference
VDTs	up - 2.8×10^{-4}	0.3	Stuchly et al. (1983)
Welding arcs (0 - 50 Hz)	0.1 - 5.8	0 - 0.8	Lövsund et al. (1982)
Induction heaters (50 - 10 Hz)	0.9 - 65	0.1 - 1	Lövsund et al. (1982)
50-Hz Ladle furnace	0.2 - 8	0.5 - 1	Lövsund et al. (1982)
50-Hz Arc furnace	up - 1	2	Lövsund et al. (1982)
10-Hz Induction stirrer	0.2 - 0.3	2	Lövsund et al. (1982)
50-Hz Electroslag welding	0.5 - 1.7	0.2 - 0.9	Lövsund et al. (1982)
Electrolyte process (0 - 50 Hz)	7.6 (mean)	operator position	Marsh et al. (1982)
Isotope separation (static fields)	1 - 50	operator position	Tenforde (1986c)

ladle furnaces (50 Hz, 13 - 15 kA) produced fields up to 10 mT, at distances up to 1 m. In the production of aluminium using a Soderberg cell, the final reduction process may lead to static field exposures of about 40 mT.

In the course of studies on the health of workers in industries using electrolytic processes, Marsh et al. (1982) found that the mean static magnetic field level at operator-accessible locations was 7.6 mT and the maximum was 14.6 mT. Time-weighted-average field exposures were calculated to be about 4 and 11.8 mT for the mean and maximum field levels, respectively.

Vyalov (1974) characterized the average magnetic field levels to which Soviet workers in permanent magnet production plants were exposed. He found that the static magnetic field at the level of a worker's hands was typically 2 - 5 mT. At the level of the chest and head, the field was generally in the range of 0.3 - 0.5 mT.

3.2.2.2 Energy technologies

High static magnetic field strengths may be encountered around new and developing technologies used for energy production and storage, such as magnetohydrodynamic systems, superconducting magnetic energy storage systems, and thermonuclear fusion (Tenforde, 1986b).

The thermonuclear fusion process involves the combination of two light nuclei to form a heavier nucleus with a resultant release of energy. Various methods can be used to confine an ignited plasma, including high-intensity magnetic fields. It is now generally believed that fields as high as 9 - 12 T will be required for the sustained magnetic confinement of an ignited plasma. Fringe fields up to 50 mT will exist at locations within the main reactor building in areas accessible to operations personnel. Although only a limited number of scientists and maintenance personnel would normally be expected to enter fields of this intensity, it is expected that they will do so for brief periods during normal reactor operation.

Power generation by magnetohydrodynamic (MHD) separation of ionic charges has been studied as a potential means for increasing the net power output of a gas- or coal-fired electric power facility. To a first approximation, a typical MHD generator can be represented as a magnetic dipole with a large net moment of approximately 8000 $MA.m^2$ (Hassenzahl et al., 1978). The field level at a distance of about 50 m from the device would then be approximately 10 mT and the field level would fall below 0.1 mT only at distances greater than 250 m.

3.2.2.3 Switching stations and power plants

Typical values for the magnetic flux density at workplaces, near overhead lines, in substations, and in power stations (16, 2/3, 50, 60Hz) range up to 0.05 mT (Krause, 1986).

3.2.2.4 Research facilities

Selected groups of workers in research laboratories may be exposed to high magnetic field strengths, particularly near bubble chambers and particle accelerators.

During the last three decades, bubble chambers have played a major role in the study of high-energy nuclear reactions. The bubble chamber is contained within a solenoidal magnet operating at field levels up to approximately 3 T.

At the location where an operator changes the film cassettes, the field is estimated to be approximately 0.4 -

0.5 T at foot level and about 0.05 T at the level of the head. The film changing procedure requires 5 min to complete, and is carried out approximately three times per day, i.e., once per 8-h work shift (Tenforde, 1986b).

Linear accelerators and synchrotrons have found applications in nearly every scientific field, including such areas as high-energy physics, nuclear chemistry, cancer radiotherapy, and isotope production for research and medicine. The scale of these devices ranges from a few metres to several kilometres. Similarly, the focusing and beam extraction magnets used in various accelerator designs differ widely in field strengths and in the magnetic field profile. Although high magnetic fields may be present near accelerator magnets, personnel are seldom exposed to these fields, because of exclusion from the high ionizing radiation zone surrounding the beam line.

3.2.2.5 Video display terminals

The use of computers with screen-based output units or video display terminals (VDT) grows at an ever increasing rate. VDT operators have expressed concerns about possible effects from emissions of low-level radiations. Magnetic fields (frequency 15 - 125 kHz) as high as 0.69 A/m (0.9 μT) have been measured close to the surface of the screen (Bureau of Radiological Health, 1981) under worst-case conditions. This result has been confirmed by many surveys (Roy et al., 1984; Repacholi, 1985a). In a comprehensive review of measurements and surveys of VDTs by national agencies and individual experts, it was concluded that there are no radiation emissions from VDTs that would have any consequences for health (Repacholi, 1985a). There is no need to perform routine radiation measurements since, even under worst-case conditions, the emissions are well below any international or national standards.

3.3 Magnetic Fields in Medical Practice

3.3.1 Diagnosis, magnetic resonance imaging, and metabolic studies

Magnetic resonance (MR) imaging used for diagnostic purposes involves both static and time-varying magnetic fields. MR imaging applied to living tissues provides a promising new technique for medical imaging with high spatial resolutions (Budinger & Lauterbur, 1984). In this technique, nuclear magnetic moments are aligned by the application of a static magnetic field (B_0), and undergo a precessional motion around the field direction with a Larmor frequency characteristic of each nucleus (section 2.3.3). When a radiofrequency (RF) field with a matching frequency is applied transverse to the direction of B_o, a resonant energy absorption occurs. The return of the magnetic spin state to equilibrium following resonant energy absorption is characterized by two relaxation times, T_1 and T_2. The T_1 parameter is called the spin-lattice relaxation time, and reflects the local temperature and viscosity in the vicinity of the magnetic nuclei. The T_2 parameter is called the "spin-spin" relaxation time, and reflects the local magnetic field resulting from the nuclear moments of neighbouring nuclei. Both the T_1 and T_2 relaxation times provide information that can be converted into contrast differences in NMR images of tissue-proton density. The intensity of the radiated signal reflects the tissue concentration of magnetic nuclei such as protons, ^{13}C, ^{23}Na, ^{31}P, and ^{39}K. The selective detection of different magnetic nuclei is possible, because of their different characteristic resonant frequencies at a given magnetic field strength.

The decay of a MR signal occurs with a characteristic time variation that conveys detailed information about the local environment of the magnetic nuclei. In proton MR images, large contrast differences can be observed between regions of tissue that have significantly different water or lipid contents. Various MR imaging methods have been developed that are able to demonstrate differences between normal and pathological regions of the same tissue (Crooks & Kaufman, 1983). Principles and applications of magnetic resonance techniques in medicine can be found in Mansfield & Morris (1982), Foster (1984), and Mathur (1984).

In addition to use as an imaging technique, MR spectroscopy based on ^{13}C and ^{31}P signals can provide unique information on tissue metabolism. For example, ^{31}P MR spectroscopy has been shown to give quantitative information on phosphate metabolism in the heart, liver, kidney, brain, and muscle tissue.

The present generation of MR imaging devices, used in clinical practice, employ stationary magnetic fields with intensities ranging from 0.3 T to about 2 T and RF fields with frequencies up to 100 MHz (the proton resonant frequency in a 2 T field is 85.15 MHz). In addition, weak spatial gradients of the stationary magnetic field (about 0.001 T/m) are used to define the tissue location of MR signals. The gradient direction is rapidly switched from one projection axis to the next in order to reconstruct the entire image of the specimen. These rapidly switched gradient fields produce a time-varying magnetic field within the tissue volume. In the MR imaging devices that are currently in existence, the maximum time rate of change of the magnetic field is normally about

1.5 T/second, but may be considerably higher in a few specialized devices.

The feasibility of using static magnetic fields with strengths greater than 2T is being explored as a means of increasing the signal-to-noise ratio in MR images. In addition, the use of higher fields could significantly reduce the time required to obtain chemical shift images, which provide high-resolution information on the spatial distribution of ^{31}P nuclei and protons associated with tissue water and fat.

3.3.2 Therapy

Patients suffering from bone fractures that do not heal well or unite have been treated with pulsed magnetic fields (Bassett et al., 1974, 1977, 1982; Mitbreit & Manyachin, 1984). Studies are also being conducted on the use of pulsed magnetic fields to enhance wound healing and tissue regeneration.

Various devices generating magnetic field pulses are used for bone growth stimulation. A typical example is the device that generates an average magnetic flux density of about 0.3 mT, a peak strength of about 2.5 mT, and induces peak electric field strengths in the bone in the range of 0.075 - 0.175 V/m (Bassett et al., 1974). Two different pulse patterns are used: a quasi-rectangular pulse of 250 - 400 µs duration with a secondary pulse of opposite polarity of 20 µs width, and a repetition rate of 40 - 77 Hz; and a train of pulses with a duration of 2.5 ms and a repetition rate of 5 - 20 Hz (Fig. 3). Near the surface of the exposed limb, the device produces a peak magnetic flux density of the order of 1.0 mT causing peak ionic current densities of about 10 to 100 mA/m^2 (1 to 10 $\mu A/cm^2$) in tissue. These ionic currents perturb cell function, even though most of the current flows around the cell in the extracellular space (Pilla, 1979; Beltrame et al., 1980; Pilla et al., 1983). Applications of magnetic field devices in medicine are rapidly expanding. Further information can be obtained in the monograph edited by Bistolfi (1983). Magnetic fields are being widely used in the USSR for various therapeutic applications (Bogolyubov, 1981).

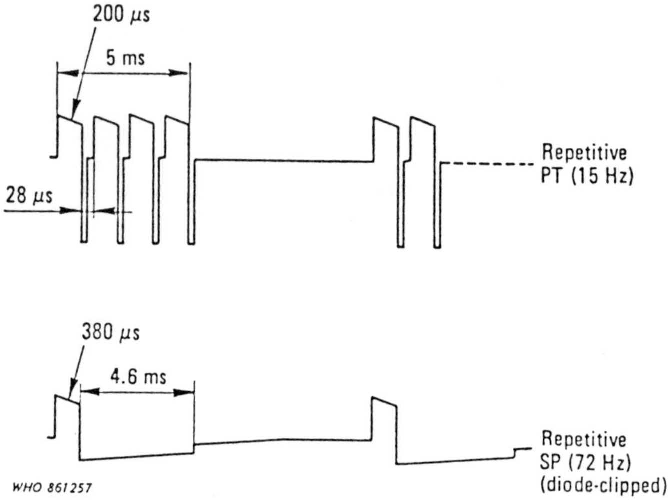

Fig. 3. Time-varying pulsed magnetic fields (dB/dt) in bone-growth stimulation and fracture repair. PT = pulse train; SP = single pulse. From: Bogolyubov (1981).

4. MECHANISMS OF INTERACTION

A broad spectrum of interaction mechanisms can occur between magnetic fields and living tissue. At the level of macromolecules and larger structures, interactions of stationary magnetic fields with biological systems can be characterized as electrodynamic or magnetomechanical in nature. Electrodynamic effects originate through the interaction of magnetic fields with electrolyte flows, leading to the induction of electrical potentials and currents. Magnetomechanical phenomena include orientational effects on macromolecular assemblies in homogeneous fields, and the translation of paramagnetic and ferromagnetic molecular species in strong gradient fields. Magnetic fields that are time-varying also interact with living tissues at the macroscopic and microscopic levels to produce circulating currents via the mechanism of magnetic induction. The theory behind each of these interaction mechanisms will be described in this section.

At the atomic and subatomic levels, several types of magnetic field interactions have been shown to occur in biological systems (Cope, 1971, 1973, 1978, 1981). Two such interactions are the nuclear magnetic resonance in living tissues described earlier and the effects on electronic spin states and their relevance to certain classes of electron transfer reactions described in this section.

Other interaction mechanisms that are being studied at the present time are discussed at the end of this section. Recent reviews of the theoretical bases for magnetic field interactions include those of Bernhardt (1979, 1986), Schulten (1982, 1986), Pirusyan & Kuznetsov (1983), Abashin & Yevtushenko (1984), Swicord (1985); Tenforde (1985a,c, 1986a,d), Kaune (1985), Frankel (1986), and Tenforde & Budinger (1986).

4.1 Static Magnetic Fields

4.1.1 Electrodynamic and magnetohydrodynamic interactions

Steady flows of ionic currents interact with applied stationary magnetic fields via the well known Lorentz force law (equation 3):

$$F = q\ (\mathbf{v} \times \mathbf{B}) \qquad (3)$$

where F is the net force exerted on a charge q moving with velocity **v**, and B is the magnetic flux density. The term **v x B** represents a vector cross-product. In the case of electrolytes flowing through channels (e.g., blood vessels), the interaction of an applied magnetic field with ionic charge carriers under steady-state conditions will result in a local force on the charge carriers of magnitude q **v** B sinΘ where Θ is the angle between the direction of charge motion and the magnetic field. This force will be perpendicular to both the magnetic field and the direction of current flow, i.e., the induced field, E_i, is transverse to both **v** and **B**. This phenomenon, which is the basis of the Hall effect in solid state materials, is also relevant to biological processes that involve electrolyte flow.

An interesting example of the role of magnetically-induced electrical potentials in a biological system is the geomagnetic direction-finding mechanism used by elasmobranch fish, including the shark, skate, and ray (Kalmijn, 1974, 1978, 1981, 1984; Ilinsky & Brown, 1985). The heads of these animals contain long jelly-filled canals known as the ampullae of Lorenzini, which have a high electrical conductivity similar to that of seawater. As the fish swims through the earth's magnetic field, a small voltage gradient is induced in the canals, which is detected by the sensory epithelia lining the terminal ampullary region. The induced electric field, which can be detected at levels as low as 0.5 μV/m (Kalmijn, 1982), has a distinct polarity that is dependent on the relative orientation of the geomagnetic field direction of swimming. In this way, the marine elasmobranchs use the **-(v x B)** fields induced in their ampullary canals as a directional compass.

A second example of induced electric potentials is provided by blood flow in the presence of an applied static magnetic field. For the specific case of a cylindrical vessel with a diameter (d) and the local electric field strength ($\mathbf{E}_i$), the magnitude of the induced potential Ψ, is given by equation 4:

$$\Psi = |\mathbf{E}_i| d = |\mathbf{v}||B| \, d \sin\Theta \qquad (4)$$

The existence of magnetically-induced blood flow potentials in the central circulatory systems of several species of mammals has been demonstrated experimentally. These induced potentials can be conveniently studied from electrocardiogram (ECG) records obtained with surface electrodes. The ECG signal in the T-wave region shows a substantial augmentation in the presence of magnetic fields and this phenomenon is completely and immediately reversible on termination of the exposure. Based on its temporal sequence in the ECG record, the increased amplitude of the T-wave in magnetic fields has been attributed to the superposition of an induced potential associated with pulsatile blood flow into the aortic vessel. This effect is illustrated in Fig. 4 and discussed in more

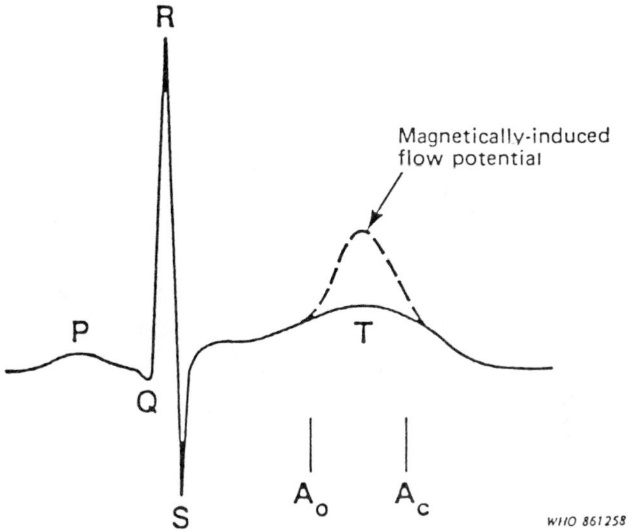

Fig. 4. Schematic diagram of the aortic blood-flow potential induced by an external magnetic field and superimposed on the normal electrocardiogram. A_o and A_c denote the approximate time at which the aortic valve opens and closes. From: Tenforde (1984).

detail in section 5. The occurrence of a change in the ECG is an excellent example of a physical effect of an applied magnetic field that does not result from a biological response to the field.

For a man with a peak blood flow rate of 0.63 m/s and an aortic diameter of 0.025 m, the predicted maximum value of the aortic flow potential is 16 mV per tesla (Mansfield & Morris, 1982). The actual potential across the cardiac muscle fibres would be much smaller, so that the threshold change in cardiac potential required to initiate depolarization of cardiac muscle may not be reached, even in magnetic fields of a few tesla. However, the induced potential differences can be significant in cases where the excitation stimulation or conduction of excitation is impaired. This is one of the reasons why, in recommendations for the safe medical use of magnetic resonance equipment where a field of more than 2 T is used, monitoring of cardiac and circulatory function of the patient is recommended (Bernhardt & Kossel, 1984, 1985).

Potential differences may also be induced by moving cross sections in a magnetic field, e.g. by cardiac contractions. In theory (Bernhardt & Kossel, 1984), a field strength of 0.1 V/m or a current density of about 10 - 20 mA/m^2 per tesla is induced in cardiac muscle. A large current density in the vicinity of the heart may cause ventricular fibrillation. However, the threshold magnetic field strengths for the induction of effects on cardiac function, including alteration of excitation or impulse conduction, are not known.

Another biological process involving ionic flows that are subject to electrodynamic interaction with an applied magnetic field is the conduction of electrical impulses in nerve tissue. Wikswo & Barach (1980) have calculated that a magnetic field strength of 24 T could produce a deflecting force on nerve ionic currents equal to one tenth of the force that they experience from interaction with the electric field of the nerve membrane. A theoretical model suggests that magnetic fields with flux densities of 2 T or less should not produce any measurable change in the conduction velocity of nerve impulses. This conclusion is supported by experimental data (section 5).

Theoretical analyses of magnetic field interactions with nerve ionic currents have also been made by Valentinuzzi (1965) and Liboff (1980). Liboff has raised the interesting question of whether time variations in the magnetic flux linkage with ion current loops along the nerve membrane could lead to significant induced potentials. Due to the rotational symmetry of the nerve axon, it is expected that these induced electrical fields would cancel. However, for the unlikely condition of highly asymmetric current loops, Liboff (1980) suggests that applied fields of less than 1 T could theoretically introduce significant perturbations in the membrane current flows during impulse conduction. At present, there are not sufficient data to test this hypothesis.

Theoretically, intracellular ionic fluxes are also susceptible to magnetic field interaction, but there is little experimental information relating to this possibility (Czerski, 1986).

4.1.2 Magnetomechanical effects

4.1.2.1 Orientation of diamagnetically anisotropic macromolecules

A large number of diamagnetic biological macromolecules exhibit orientation in strong magnetic fields. In general, these macromolecules have a rod-like shape, and magneto-orientation occurs as a result of an anisotropy in the magnetic susceptibility tensor (χ) along the different axes of rotational symmetry. The magnetic moment per unit volume (M) of these molecules in a field with intensity H is equal to χH. The theoretical calculation of the interaction energy per unit volume has been discussed by Tenforde

(1985a) and Frankel (1986). The rod-like molecules will rotate to achieve a minimum energy in the applied magnetic field. For individual macromolecules, the magnetic interaction energy predicted theoretically will be small compared to the thermal interaction energy kT, unless enormous field strengths are used. This fact has been demonstrated for DNA solutions in which the extent of magneto-orientation has been studied from measurements of magnetically-induced birefringence (the Cotton-Mouton effect). Measurements on calf thymus DNA (Maret et al., 1975; Maret & Dransfeld, 1977), resulted in a degree of orientation of only 1% in an applied field of 13 T.

Despite the weak interaction of individual macromolecules with intense magnetic fields, there are several examples of macromolecular assemblies that exhibit orientation in fields of 1 T or less. This phenomenon results from a summation of the diamagnetic anisotropies of the individual molecules within the assembly, thereby giving rise to a large effective anisotropy and magnetic interaction energy for the entire molecular aggregate. Examples of biological systems that exhibit orientation in fields of 1 T or less are retinal rod outer segments (Chalazonitis et al., 1970; Hong et al., 1971; Vilenchik, 1982), photosynthetic systems such as chloroplast grana, photosynthetic bacteria, and Chlorella cells (Geacintov et al., 1971, 1972; Becker et al., 1973, 1978a,b; Breton, 1974), purple membranes of Halobacteria (Neugebauer et al., 1977), muscle fibres (Arnold et al., 1958), and "sickled" erythrocytes (Murayama, 1965). A more detailed discussion can be found in Maret & Dransfield (1985).

Several of the physical principles underlying magneto-orientation phenomena have been experimentally demonstrated for retinal rod outer segments. The first observation that isolated rod outer segments, which consist of pigmented disc membranes stacked in a regular array, will orient in a 1 T stationary magnetic field was made in 1970 (Chalazonitis et al.,. 1970). The oriented segments are aligned with the disc membranes perpendicular to the applied field direction, which indicates that magneto-orientation results from the large summed diamagnetic anisotropy of the rhodopsin photopigments, as opposed to the lamellar membrane phospholipids (Hong, 1980; Becker et al., 1978b). An estimate of the summed anisotropy for the rod outer segments can be obtained by observing the kinetics of the magneto-orientation process. The time for rods to rotate by 90° is predicted to be approximately 4 seconds in a 1.0 T field (Hong et al., 1971; Hong, 1977, 1980), and this value agrees well with experimental observations on the kinetics of rod orientation (Chagneux & Chalazonitis, 1972; Chagneux et al., 1977). It should be noted that the slow orientational response of rod outer segments to an applied magnetic field makes this mechanism an unlikely basis for the magnetophosphene phenomenon that is observed in time-varying fields.

4.1.2.2 Orientation of organisms with permanent magnetic moments

An interesting specimen for the biophysical study of magnetic field interactions was provided by Blakemore's accidental discovery of magnetotactic bacteria (Blakemore, 1975). Approximately 2% of the dry mass of these aquatic organisms is iron, which has been shown by Mossbauer spectroscopy to be predominantly in the form of magnetite (Fe_3O_4) (Frankel et al., 1979). The magnetite inclusions are arranged as chains of approximately 20 - 30 single domain crystals. The orientation of the net magnetic moment is such that magnetotactic bacteria in the northern hemisphere migrate towards the north pole of the geomagnetic field, whereas strains of these bacteria that grow in the southern hemisphere move towards the south magnetic pole (Blakemore et al., 1980; Rosenblatt et al., 1982a,b). Magnetotactic bacteria that have been found at the geomagnetic equator are nearly equal mixtures of south-seeking and north-seeking organisms (Frankel et al., 1981). Because of the polarities of their magnetic moments, the magnetotactic bacteria in both the northern and southern hemispheres migrate downwards in response to the vertical component of the geomagnetic field. It has been proposed that this downward directed motion, which carries the bacteria into the bottom sediments of their aquatic environment, is essential for the survival of these microaerophilic organisms (Blakemore, 1975; Frankel et al., 1979). This phenomenon is an interesting example of an interaction between a physical response to a magnetic field and biological environmental adaptation and selection processes.

4.1.2.3 Translation of substances in a magnetic field gradient

A material with a net magnetic moment will experience a force in a magnetic field gradient (spatially non-uniform magnetic field). As a result of this force, paramagnetic and ferromagnetic materials will migrate along the direction of the magnetic field gradient.

One of the interesting applications of the magnetomechanical force exerted by a magnetic field gradient is the differential separation of erythrocytes from whole blood (Melville et al., 1975; Paul et al., 1978). In this procedure, deoxygenated erythrocytes, in which the haemoglobin is paramagnetic, are attracted to a wire mesh with a strong gradient field and thereby separated from other classes of blood cells. Magneto-mechanical forces are also applied in: surgical traumatology to fix skeleton elements in specific position (Yarovitsky, 1986); in designs of various intestinal and oesophageal valves; and for the accumulation of drugs (in compounds with ferromagnetics) in specific parts of the human body. In ophthalmology, strong constant magnets are applied to extract foreign (ferromagnetic) objects.

An important safety consideration is the displacement of metallic inclusions or implants in human beings exposed to strong magnetic field gradients, as this could pose a health risk (New et al., 1983).

4.1.3 Effects on electronic spin states

A number of organic reaction processes that involve electron transfer via radical pair intermediates are highly sensitive to magnetic field interactions. A well-studied example that is biologically relevant is the photo-induced charge transfer reaction that occurs in bacterial photosynthesis (Blankenship et al., 1977; Werner et al., 1978; Haberhorn & Michel-Beyerle, 1979; Michel-Beyerle et al., 1979; Hoff, 1981; Ogrodnik et al., 1982). Within 10 picoseconds (ps) following excitation of bacteriochlorophyll (BChl) to its first excited singlet state, a radical pair intermediate state is formed that consists of a $(BChl)^{+}_{2}$ cationic dimer and a bacteriopheophytin $(BPh)^{-}$ anion. Within 200 ps, electron transfer occurs to the ultimate acceptor, an ubiquinone-iron complex. However, if the acceptor molecule is chemically reduced, the lifetime of the radical pair intermediate state increases to approximately 10 nanoseconds (ns). With an extended lifetime, hyperfine interactions between the nuclear and electron spin magnetic moments lead to an interconversion of the radical pairs between the singlet and triplet states. Under this condition, the intermediate state decays directly back to the singlet ground state, or decays via a metastable triplet state. Because of the weakness of the hyperfine interaction, the triplet states are nearly degenerate and the electron spins of the radical pair intermediate can move with nearly equal probabilities between the singlet S_o and the triplet T_o and $T_{\pm 1}$ states. However, in the presence of an applied magnetic field that exceeds approximately 10 mT, the resulting Zeeman interaction with the radical electron spins will lift the degeneracy of the triplet state and effectively block the $T_{\pm 1}$ triplet channels. Theoretically, the yield of triplet product should be reduced by two thirds in the presence of the external field, and this has been confirmed experimentally by laser pulse excitation and optical absorption measurements (Michel-Beyerle et al., 1979).

In considering the biological implication of these studies, it should be kept in mind that the ultimate electron acceptor molecules have been altered by chemical reduction and such conditions do not normally occur in nature. However, the possibility cannot be excluded that similar phenomena may occur in other radical-mediated biological processes under normal conditions. It has been proposed, for example, by Schulten et al. (1978), that an anisotropic Zeeman interaction with a radical mediated reaction system could provide a basis for geomagnetic direction finding.

Magnetic field effects on organic chemical reactions in which the splitting and subsequent recombination of a non-excited singlet molecule involves a radical pair as a short-lived intermediate stage have also been well documented (Molin et al., 1979; McLauchlan, 1981). As described above, the effect of the magnetic field is on the singlet-triplet transition rate of the radical pair, thus affecting the relative proportion of recombinant and escape products, by up to 30% in some cases. The magnitude of the response depends on the difference in the magnetic properties of the two radical intermediates, and will be particularly enhanced in reactions involving transition metals such as iron (Molin et al., 1979). These effects also increase with the lifetime of the radical pair. This is typically 100 ps - 100 ns in solution, but is longer when the reacting molecule is held in a micellar "cage" or bound to an enzyme.

A number of enzyme reactions involving radical intermediates have been identified (Saunders & Cass, 1983), though the evidence is tenuous because they are generated and react at the enzyme active site and their presence can only be inferred by indirect methods. Enzymes, the action of which may involve radical intermediates, are:

(a) Cytochrome P-450: A class of haem-containing enzyme involved in drug metabolism and steroid hydroxylation;

(b) Lipoxygenase: A non-haem iron enzyme that is a key enzyme in prostaglandin and thromboxane synthesis; and

(c) Cyclo-oxygenase: The enzyme involved in converting arachadonic acid to prostaglandins.

These enzymes all contain iron and use oxygen (O_2) as one of the substrates. They can be expected to be sensitive to magnetic fields, if the radical recombination is rate determining, i.e., if the radical is relatively long-lived.

4.2 Time-Varying Magnetic Fields

In accordance with Faraday's law, magnetic fields that vary in time will induce potentials and circulating currents in biological systems.

$$J = E\sigma = \frac{\pi r^2}{2\pi r} \times \frac{dB}{dt} \times \sigma = \frac{\sigma r}{2} \frac{dB}{dt} \qquad (5)$$

where J = current density (A/m^2)
E = induced potential (V/m)
r = radius of the inductive loop (m)
σ = tissue conductivity (S/m)
$\frac{dB}{dt}$ = rate of change of magnetic flux density

For sinusoidal fields of frequency f, equation (5) reduces to:

$$J = \pi r f \sigma B_o,$$

where B_o is the magnetic field amplitude.

Thus, the magnitude of the induced electric fields and current densities is proportional to the radius of the loop, the tissue conductivity, and the rate of change of magnetic flux density.

The dependence of the induced field and current on the radius of the loop through which magnetic flux linkage occurs is an important consideration for biological systems. Time-varying fields of modest strength may induce significant circulating currents at the macroscopic level, but substantially smaller currents at the cellular level.

An important factor to be considered in the response of biological systems to a time-varying magnetic field is the waveform. Many different types of magnetic field waveform are used in practice, including sinusoidal, square-wave, saw tooth, and pulsed fields. For these fields, the two parameters of key importance are the rise and decay times of the signal, which determine the maximum time rates of change of the field, (dB/dt), and hence the maximum instantaneous current densities induced in tissues. These also depend on tissue conductivity, which is frequency dependent and differs between tissues.

Luben et al. (1982) and Cain et al. (1984) demonstrated *in vitro* that pulsed magnetic fields, generated in pulse trains (72 Hz) or recurrent bursts (15 Hz), blocked the response of mouse osteoblasts to parathyroid hormone. The effects seemed to be mediated at the cell membrane by blocking receptor-adenylate cyclase coupling in the membrane (Cain et al., 1985). The adenylate cyclase and cyclic AMP systems are part of the hormone response amplification system. These effects were associated with current densities of 10 - 100 mA/m^2 and electric field strengths of 0.1 - 1 V/m in extracellular fluids. Effects at the cell membrane receptor level seem to be involved in the effects of 450 MHz fields sinusoidally modulated at ELF frequencies on T-lymphocyte cytotoxic functions (Lyle et al., 1983).

Numerous effects resulting ELF electric fields in cells and tissues, induced by pulsed magnetic fields, have been described. These effects include stimulation of bone growth, nerve and limb regeneration, cell differentiation, effects on ionic fluxes and on DNA, RNA, and protein synthesis (Sheppard, 1985). Experimental data seem to indicate that the site of the primary interaction is the cell membrane and proposed mechanisms presume a role of the induced electric field. Effects on gene expression, such as the initiation and alteration of transcription (Goodman et al., 1983; Goodman & Henderson, 1986), or effects on Escherichia coli lac operon (Aarholt et al., 1982) may be mediated through interaction with the genetic apparatus (chromosomes). Effects on cell membrane and/or gene expression may be responsible for abnormalities in chick embryo development described by Delgado et al. (1982) and Ubeda et al. (1983). However, the latter studies have not been confirmed by Maffeo et al. (1984).

A well-documented biological effect of time-varying magnetic fields is the occurrence of magnetophosphenes. Various investigations leading to the elucidation of this phenomenon are summarized in Table 7. First observed by d'Arsonval (1896), magnetophosphenes are detected as a sensation of flickering light induced in the eye, when it is exposed to magnetic fields with flux densities greater than about 10 mT and frequencies greater than 10 Hz. The minimum field strength required to produce visual phosphenes (Fig. 5) occurs at a frequency of 20 Hz (Barlow et al., 1947b; Lövsund et al., 1979, 1980a,b; Tenforde & Budinger, 1986). There is evidence (Lövsund et al., 1981) to suggest that the time-varying magnetic field effect occurs in the photoreceptors rather than in the post-synaptic neurons. Furthermore, Lövsund et al. (1980a,b, 1981) concluded from their studies on volunteers that the mechanisms of underlying magnetically and electrically-induced phosphenes are possibly the same.

The idea that the magnetically-induced electric field strength (and hence current density) in tissues is the physical quantity determining the biological effects at the cellular level has been pursued by Bernhardt (1979, 1985). He used electrophysiological data in order to find "safe" and "hazardous" current densities and to define corresponding magnetic field strengths. The problem is the correlation of the internal current densities with the external magnetic field strengths. Bernhardt (1985) concluded that, using his calculations, it was possible to estimate the current density

Table 7. Magnetophosphene studies

Reference	Principal findings
d'Arsonval (1896)	Initial report of magnetophosphenes produced by a 42-Hz field
Thompson (1909-10)	Described magnetophosphenes produced by a 50-Hz field as a colourless, flickering illumination that is most intense in the peripheral region of the eye
Dunlap (1911)	Demonstrated that magnetophosphenes produced by a 25-Hz field are more intense than those produced by a 60-Hz field of comparable intensity
Magnusson & Stevens (1911-12)	Demonstrated the production of magnetophosphenes by pulsed DC fields as well as by time-varying fields with frequencies from 7 to 66 Hz; observed strongest magnetophosphenes with fields oscillating at 20 - 30 Hz
Barlow et al. (1947a,b)	Demonstrated threshold field intensity of 20 mT (rms) at 30 Hz, and showed that the threshold for magnetophosphenes is relatively insensitive to background illumination compared with that for electrophosphenes; characterized "fatigue" phenomenon with a 60 Hz magnetic field applied for 1 min, which was followed by a refractory period of 40 s, during which a second phosphene could not be elicited; demonstrated that magnetic fields must be applied in the region of the eye to produce phosphenes, and that sensitivity is abolished by pressure applied to the eyeball
Seidel et al. (1968)	Observed comparable light patterns associated with visual stimulation by ELF electric and magnetic fields, but found different probabilities of occurrence of certain types of phosphene patterns
Lövsund et al. (1979-81)	Analysed threshold field intensity for production of magnetophosphenes over frequency range of 10 - 45 Hz; demonstrated maximum sensitivity to a 20-Hz field; studied effects of dark adaptation, background illumination, and visual defects on sensitivity to magnetophosphenes; compared threshold stimuli required to produce electrophosphenes and magnetophosphenes; characterized changes in electrophysiological responses of isolated frog retinas exposed to ELF magnetic fields
Silny (1981); Bernhardt (1985)	Found minimum time rate of change for magnetophosphenes in sinusoidal fields at 17 Hz to be 0.3 T/s
Budinger et al. (1984a)	Found minimum time rate of change of pulsed magnetic field to be 1.3 - 1.9 T/s to produce magnetophosphenes

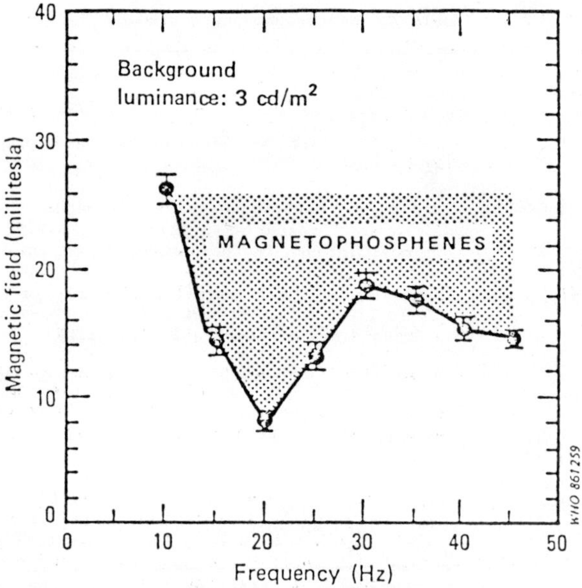

Fig. 5. Frequency dependence of threshold values for magnetophosphene induction. From: Lövsund et al. (1980a). At 20 Hz (the frequency of maximum sensitivity), the time rate of change of the magnetic field (dB/dt) is 1.3 T/s.

levels within one order of magnitude. Satisfactory agreement of theoretical predictions with data on the threshold for magnetophosphene perception in human volunteers was obtained (Bernhardt, 1985). This was established to be between 2 and 10 mT (for frequencies greater than 10 Hz).

Another potentially important target of ELF magnetic field interactions is the nervous system. From a consideration of the naturally occurring fields in the central nervous system, Bernhardt (1979) concluded that magnetic fields in the 1 - 100 Hz frequency range, which can induce current densities in tissue of approximately 1 mA/m² or less, should not have a direct effect on the brain's electrical activity. The strength of a 60-Hz magnetic field that would induce a peak current density of this magnitude in the cranium of a human subject was calculated to be about 0.5 mT (Tenforde, 1985a).

In a careful study on human perception to 60 Hz magnetic fields, Tucker & Schmitt (1978) did not find any significantly perceptive individuals among more than 200 subjects exposed to a field with an amplitude of 2.1 mT. Several behavioural tests on mice exposed to a 60-Hz magnetic field that induced a peak current density approaching 1 mA/m² in the peripheral cranial region also yielded negative findings (Davis et al., 1984). The results of these studies suggest that ELF magnetic fields must have significantly greater amplitudes than the theoretically calculated threshold values in order to perturb animal behaviour. However, it is important to recognize the inherent deficiencies of a simple theoretical model that treats the central nervous system as a region of uniform conductivity. In addition, the induced current in a loop of maximum radius at the brain's surface may not be the relevant parameter to consider in predicting the response to ELF magnetic fields. The regions of the central nervous system that might be responsive to these fields may have significantly smaller dimensions than the entire cranium. Thus, a large increase in the ELF magnetic field strength would be necessary to evoke a measurable electrical and/or behavioural perturbation.

It is reasonable to suppose that these effects result from the interaction of the induced electric fields and currents with the membranes of nerve and muscle cells, thereby causing changes in the electrical excitability of these cells in the same way as naturally occurring or directly applied electric fields.

The permeability to ions of the nerve (and muscle) cell membranes depends on the membrane potential. It is this voltage-dependent permeability that gives the cells the property of being electrically excitable. When an electric field is applied, various charged side-groups of certain proteins embedded in the membrane change their configuration, thereby causing a larger structural change in the protein as a whole. In this new conformation, ions are able to pass through the membrane by binding temporarily with the protein molecule at various sites, thus "hopping" through the membrane. In any area of a membrane, there are a large number of "gating" molecules, and the effect of an induced electric field may involve an alteration in the proportion of gates that are open. This type of interaction could significantly influence membrane permeability.

In addition, there is a specific type of protein molecule for each species of ion, permitting different ionic responses to the same electric field. Thus, in response to a depolarizing electric field, there is a large increase in Na^+ permeability in the membrane of a nerve cell tending to depolarize the cell further. This event is followed by a slower change in K^+ permeability and an inactivation of the Na^+ channel, resulting in a repolarization. Induced fields sufficient to exceed a threshold depolarization value can result in an action potential that is capable of stimulating other excitable cells. These effects are well understood.

ELF magnetic fields inducing such large depolarizations may result in nerve stimulation or muscle contraction, or even in fibrillation. ELF magnetic fields inducing weak electric fields may also interact with, or modulate, nervous system activity in a manner that is less well understood. However, these interactions can produce changes in electrical excitability. Such interactions may be involved in, for example, magneto- or electrophosphenes.

The ELF field interactions described above exhibit frequency-dependent thresholds characteristic of nervous tissue, and have been well documented by Bernhardt (1979, 1985). This frequency dependence is very important when relating experimental results obtained using high frequencies or very short pulses to effects anticipated at 50/60 Hz, at least as far as acute responses are concerned. The main factors governing this dependence are accommodation and ionic mobility. As a result, there is a characteristic U-shaped dependence of threshold current density on frequency, with the lowest values for most nervous tissues occurring between 10 Hz and 100 kHz. At low frequencies, the effects of accommodation predominate, which is thought to be related to the slow inactivation of the Na^+ channel. At higher frequencies, the time available during each cycle for ions to migrate across the membrane, an all or nothing event, becomes limiting; direct electrical excitation gives way to heating somewhere between 100 Hz and 300 kHz.

It should be noted that the theoretically calculated field intensities at 20 Hz for stimulating the visual system are only slightly lower than the perception threshold for magnetophosphenes (Bernhardt, 1985). With regard to "hazardous values" and the upper limit of the field strength that leads to injury, the ultimate criterion for the definition of injury may be the initiation of heart fibrillation. The threshold for extra-systole induction at 60 Hz is estimated to be above 300 mT for stimulation times of 1 second or longer, and the threshold for ventricular fibrillation is higher by a factor of 3 - 5 (Bernhardt, 1985). For shorter exposure times, higher field strengths are necessary to produce similar biological effects.

Silny (1986) measured the stimulation threshold of the heart in 8 dogs exposed to time-varying magnetic fields. He converted the thresholds found in dogs to the equivalent thresholds expected in human beings. From his data, the fibrillation threshold for the human heart was estimated to be 1 T at 50 Hz, for magnetic fields acting perpendicular to the body axis.

4.3 Other Magnetic Field Interactions Under Study

The transduction mechanism for ELF magnetic fields described in section 4.2 is supported by experimental data for electrically excitable tissues. For other biological effects observed with ELF fields that induce smaller current densities (below the level that could significantly affect the cell membrane potential), other transduction mechanisms have been proposed. For example, changes in cell-surface receptor molecules and in ion binding to membrane surfaces have been reported to occur, as a result of exposure to ELF magnetic fields. It has been proposed that the pericellular currents induced by an ELF field may produce electrochemical alterations in components of the cell membrane surface. These events, in turn, send signals across the cell membrane barrier that produce alterations in intracellular biochemical and physiological functions. This hypothesized scheme of transductive coupling between induced electric currents in the extracellular medium and the intracellular events occurring in living cells is illustrated schematically in Fig. 6.

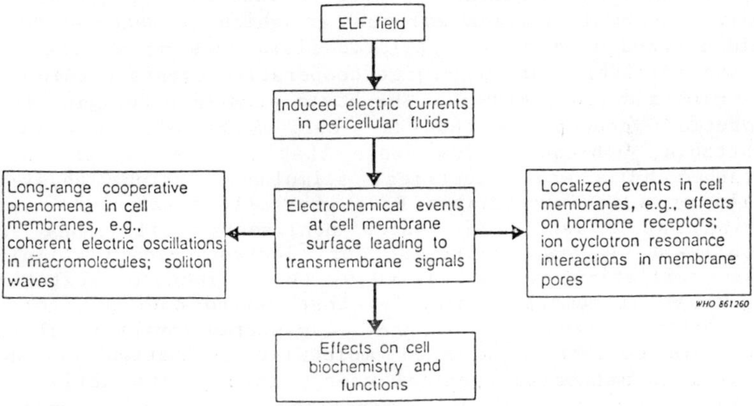

Fig. 6. Proposed mechanisms of ELF magnetic field interactions at cell surfaces. From: Tenforde & Kaune (in press).

The key element in the sequence of events through which externally applied ELF fields influence cellular properties is the transductive signalling event within the cell membrane. Numerous theoretical models have been proposed for the transmembrane signalling process (or processes) that are triggered by induced pericellular electric currents (Adey,

1980, 1981, 1983). In the broadest sense, these hypothesized mechanisms can be grouped into two general classes:

(a) long-range cooperative phenomena established within the matrix of glycoproteins and lipoproteins that constitute the cell membrane; and

(b) localized events occurring at specific ligand-binding sites (receptors) at the outer membrane surface, or events occurring within ion-selective channels that span the membrane and electrically couple the intracellular and extracellular fluids.

These classes of phenomena are depicted by the boxes at the left and right sides of Fig. 6 and will be discussed separately.

4.3.1 Long-range cooperative phenomena in cell membranes

The electric fields induced in tissue by externally-applied low-amplitude ELF electromagnetic fields are several orders of magnitude less than the voltage gradient that exists across the living cell membrane. It has therefore been proposed that the cellular response to external ELF fields may involve an amplification process in which a weak electric field induced in the extracellular fluid acts as a "trigger" for the initiation of long-range cooperative events within the cell membrane (Adey, 1981). The basic premise underlying this theoretical concept is that the cell membrane exists in a metastable, non-equilibrium state that can be significantly perturbed by a weak electrical stimulus. Various physical models of such interactions have generally treated the cell membrane as a lattice in which nonlinear oscillations are established by weak electrical (or electrochemical) stimuli. These oscillations are amplified by the collective excitation of patches of membrane molecules that extend over a significant portion of the cell surface. The stored energy resulting from this collective mode of molecular excitation is then released as metabolic chemical energy through the activation of ion pumps or enzymatic reactions within the membrane (Frohlich, 1968, 1977; Grodsky, 1976, 1977; Kaczmarek, 1977; Lawrence & Adey, 1982, 1983; Adey, 1983).

4.3.2 Localized interactions of external ELF fields with cell membrane structures

Recent experimental evidence and theoretical models have given support to the concept that the interactions of ELF electromagnetic fields with living cells occur at specific loci on the cell membrane. In many ways, this concept is more attractive than the hypothesized long-range membrane interactions described above. Apart from the abstract nature of such theories, the concept of long-range interactions that involve a large fraction of the cell membrane surface is generally feasible only for electromagnetic fields with frequencies well above the ELF range. Recent theoretical efforts have therefore focused on the possibility that weak ELF field interactions could significantly alter either ligand-receptor interactions at the membrane surface, or the transmembrane movement of electrolytes. Theoretical and experimental developments in this area include the following:

(a) Ligand-receptor interactions

Chiabrera et al. (1984) proposed a model of membrane interactions in which a microelectrophoretic motion induced in the cell membrane by weak ELF electric fields influences the average distance between charged ligands and the cell-surface receptors to which they are bound. In this theoretical model, the effect of the imposed electric field is to decrease the mean lifetime of the ligand-receptor complexes on the membrane surface. The authors propose that this effect could influence various biological phenomena such as the activation of lymphocytes by antigens and various lectins. An ELF field interaction of this type could also influence the gating mechanisms that control the membrane transport of various types of cations such as calcium.

(b) Combined static and ELF field interactions

Some experimental evidence suggests that ion cyclotron resonance effects could occur between ELF fields and static magnetic fields with intensities comparable to that of the geomagnetic field. Briefly summarized, it has been reported that magnetic resonance conditions influence the dielectric properties and growth rate of yeast cells (Jafary-Asl et al., 1982), the rate of lysozyme reaction with a cell membrane substrate (Jafary-Asl et al., 1982), the behaviour of rats in a timing discrimination task (Liboff et al., 1985), and the rate of calcium ion release from the surfaces of brain cells exposed *in vitro* to low-intensity electromagnetic fields (Blackman et al., 1985a,b). The first two of these biological effects were claimed to occur in response to conventional nuclear magnetic resonance conditions under which the static field intensity and the frequency of the electromagnetic field were related by the Larmor relationship for various nuclei, including ^{1}H, ^{23}Na, ^{31}P, ^{35}Cl, and ^{39}K (Jafary-Asl et al., 1982). In a third study (Liboff et al., 1985), reversible changes in rodent timing behaviour were observed when rats were simultaneously exposed to a horizontal 60-Hz magnetic field and a vertical magnetostatic field with a flux density of 26 μT. This combination of static field intensity and oscillating field frequency satisfies the cyclotron resonance conditions for lithium ions, which are thought to exert neuropharmacological effects. In the fourth study (Blackman et al., 1985b), a generalized relationship was derived between the biologically effective electromagnetic field frequency and the static magnetic field flux density. This relationship established a proportionality between the frequency of the oscillating field and the static magnetic field flux density multiplied by an index, $(2n + 1)$, where $n = 0$ or 1.

Liboff (1985) proposed that these weak interactions, which involve energy transfer from the external field that is 8 orders of magnitude less than the Boltzmann thermal energy (kT), could nevertheless impart kinetic energy to ions, such as calcium, moving through transmembrane channels. The theoretical argument was made by McLeod & Liboff (1986) that ion channels provide an environment in which damping effects on ion motion due to collision may be reduced relative to the high collision frequencies that exist in bulk aqueous media. Nevertheless, a simple calculation indicates that, under the various experimental conditions described above, the induced electric field within ion transport channels is of the order of 10^{-10} V/m. This field level is 2 orders of magnitude less than the Nyquist thermoelectric noise present in membrane channels (Bawin & Adey, 1976). Overall, the experimental data that suggest a possible role of cyclotron resonance effects on ion binding to membrane surfaces and on cation transport through cell membrane pores are intriguing, but there is a clear need for refinements in the theoretical description of this phenomenon and to substantiate the experimental results.

5. EXPERIMENTAL DATA ON THE BIOLOGICAL EFFECTS OF STATIC MAGNETIC FIELDS

In this section, the aim is to present a review of experimental observations on the biological effects of exposure to magnetic fields, and to relate them to data presented in the preceding section, in which the mechanisms of interaction were discussed. Empirical observations for which no theoretical explanations are available, at present, will be pointed out, and an attempt will be made to identify gaps in knowledge. The data discussed here were selected on the basis of their relevance for the assessment of health risks. Thus, many papers have been omitted from the discussion.

Several comprehensive sources of experimental data on the biological effects of magnetic fields are available. Older results have been collected in two volumes edited by Barnothy, M.F., ed. (1964, 1969) and the monograph by Kholodov (1966, 1974); more recent results can be found in the report of the American Institute of Biological Sciences (1985). Some recent reviews include those prepared by Bogolyubov (1981), Kholodov (1982), Schulten (1982), Galaktionova et al. (1985), Sidjakin (in press), Tenforde (1979, 1985a,b,c, 1986a), and Tenforde et al. (1985). Valuable information and extensive bibliographies can be found in review papers by Budinger (1979, 1981), Budinger & Collander (1983), Persson & Stahlberg (1984), and Tenforde & Budinger (1986), which address the biological effects of magnetic fields in the context of the safety of magnetic resonance imaging and in vivo spectroscopy.

All the above reviews are concerned with potential risks for human health from exposure to magnetic fields of a strength greater than that of the geomagnetic field. This document does not deal with magnetic fields of a strength below that of the geomagnetic field. However, readers interested in this aspect are referred to reviews by Nakhil'nitskaya et al. (1978) and Kopanov & Shakula (1985).

The organization of this section will follow the order of increasing biological complexity of the system studied.

5.1 Molecular Interactions

Research on magnetic field interactions with biological molecules has led to a diversity of findings as exemplified by the results of studies on various enzymes summarized in Table 8. A total of 15 reports has appeared in which the reaction rates of 17 different enzymes were studied during exposure to stationary magnetic fields over a broad range of field strengths, and with widely varying exposure times, reaction temperatures and pH levels, and conditions of field

Table 8. Magnetic field effects on enzyme systems

Enzyme	Applied field strength (tesla)	Effect on enzyme activity	Reference
Acetylcholinesterase	1.7	increase	Young (1969)
Alcohol dehydrogenase	1.4	none	Muller et al. (1971)
Aldolase	17	none	Rabinovitch et al. (1967a,b)
Asparaginase	1.7	increase	Shishlo (1974)
β-galactosidase	1	none	Thomas & Morris (1981)
Carboxydismutase	2	increase	Akoyunoglou (1964)
Catalase	6	increase	Haberditzl (1967)
Catalase	0.8	increase	Vainer et al. (1978)
Cytochrome oxidase	1.3	increase	Gorczynska et al. (1982)
DNase	0.3	increase	Komolova et al. (1972)
Glumatic dehydrogenase	7.8	decrease	Haberditzl (1967)
Histidase	1.7	decrease	Shishlo (1974)
Lactic dehydrogenase	1.4	none	Muller et al. (1971)
Peroxidase	17	none	Rabinovitch et al. (1967a,b)
RNase	17	none	Rabinovitch et al. (1967a,b)
RNase	4.8	none	Maling et al. (1965)
RNase	1.4	none	Muller et al. (1971)
RNase	0.3	none	Komolova et al. (1972)
Succinate-cytochrome-C reductase	4.8	none	Maling et al. (1965)
Trypsin	0.8	increase	Cook & Smith (1964)
Trypsin	20.8	none	Rabinovitch et al. (1967a,b)
Trypsin	10	none	Nazarova et al. (1982)
Trypsin	1.4	none	Vajda (1980)
Tryosinase	17	none	Rabinovitch et al. (1967a,b)

uniformity. Overall, 58% of the experimental studies showed no effects of the field exposure, while 33% and 8% of the tests showed increases and decreases, respectively, in the rate of enzyme reactions in the exposed samples relative to controls. As discussed earlier in section 4, in certain systems, such as enzymes that involve radical intermediate stages as part of their reaction pathways, it might be anticipated that the reaction would be sensitive to the presence of a magnetic field. However, for several other enzyme systems there is no obvious physical mechanism that could explain the observed magnetic sensitivity at the field intensities that were used. It is interesting to note, for example, that Cook & Smith (1964) found that the activity of trypsin increased by up to 23% during a 2-h exposure to a 0.8-T field, whereas Vajda (1980) and Nazarova et al. (1982) did not observe any change in enzyme activity during exposures of 2-8 h duration in a 1.4-T field. Furthermore, Nazarova et al. (1982) found that trypsin activity was not affected by a 2.5 h exposure to a 10-T field, and Rabinovitch et al. (1967a,b) did not observe any change in trypsin activity either during a 9 min exposure to a 22-T field, or following a 3.7 h pretreatment of the enzyme in a 20.8-T field.

Another aspect of the data presented in Table 8 that merits comment is the finding in two different laboratories of an increase in the reaction rate of the metalloenzyme catalase in response to exposure to a magnetic field (Haberditzl, 1967; Vainer et al., 1978). Vainer et al. (1978) reported that the reaction rate of catalase varied linearly with field between 0 and 0.8 T, increasing by 20% at 0.8 T. The action of this enzyme may involve a radical intermediate state which, as discussed in the preceeding section, might be anticipated to exhibit magnetic sensitivity. Several other biologically important enzymes that may have radical intermediate steps in their pathways include the cytochrome P-450 enzymes, which are involved in steroid hormone metabolism, and lipoxygenase and cyclo-oxygenase, both of which are involved in the synthesis of prostaglandins (Saunders & Cass, 1983). Further studies on these enzyme systems would provide useful insight into whether enzymatic pathways that involve radical intermediate states exhibit sensitivity to a stationary magnetic field, with possible consequences for cellular and tissue functions (Schulten, 1986).

A well-studied mechanism by which static magnetic fields can influence macromolecules is through a magneto-orientational effect. As discussed in section 4, this phenomenon produces measurable effects on single molecules, only at field strengths greater than 10 T. Various macromolecular assemblies, such as retinal photopigments, can be oriented in fields of less than 1 T. However, at present, there are no data suggesting that magneto-orientation of these various macromolecules exerts profound effects on vital membrane, cellular, or tissue function. For example, mammalian visual functions have been found to be unaffected by static magnetic fields up to 1.5 T (Gaffey & Tenforde, 1984).

5.2 Effects at the Cell Level

The results of a number of studies conducted in the 1960s and earlier suggested that exposure to stationary fields might lead to physiological, morphological, and growth abnormalities at the cellular level (Barnothy, M.F., 1964, 1969). Degenerative changes such as pycnosis (Mulay & Mulay, 1961; Pereira et al., 1967), decreased DNA synthesis (D'Souza et al., 1969) and growth inhibition (Gerencer et al., 1962; Butler & Dean, 1964) were noted for various types of normal and tumour cells. In contrast to these observations, a large number of more recent studies using magnetic field intensities and exposure times that were equal to or greater than those used in the earlier work have failed to produce effects on cell growth (Montgomery & Smith, 1963; Halpern & Greene, 1964; Hall et al., 1964; Rockwell, 1977; Iwasaki et al., 1978; Frazier et al., 1979, Nath et al., 1980). It is also interesting to note that early reports (Barnothy, J.M., 1964; Gross, 1964) of in vivo tumour growth inhibition by stationary magnetic fields have not been replicated in other studies (Eiselein et al., 1961; Chandra & Stefani, 1979). All of the studies mentioned above were performed under different exposure conditions and thus are difficult to compare.

Malinin et al. (1976) reported that exposure of human WI-38 fibroblasts and murine L-929 cells to a 0.5-T field for 4-8 h at 4 °K led to subsequent growth inhibition compared with controls, when the cells were thawed and cultured at 30 °C. The exposed cultures also appeared to undergo morphological transformation and to lose sensitivity to contact inhibition of cell division in long-term cultures. These observations were later shown to be the result of using unconventional culture techniques in which control cells were subcultured at 5- to 6-day intervals, while cultures grown from exposed cells were only passaged at 28- to 45-day intervals. When Frazier et al. (1979) used similar culture techniques, they were able to replicate, in unexposed cultures of WI-38 and L-929 cells, the morphological transformation that had been reported by Malinin et al. (1976) to result from magnetic field exposure. Thus, the results of Malinin et al. (1976) should be discounted in an evaluation of magnetic field effects on cell cultures.

Although the preponderance of available experimental evidence indicates that stationary magnetic fields with intensities up to 2 T exert little influence on cell growth properties, there are potential mechanisms, discussed by Tenforde (1985b), by means of which effects might occur. These include:

(a) Enzymatic pathways that contain radical intermediate stages and may be sensitive to the presence of strong magnetic fields;

(b) The redistribution of paramagnetic oxygen molecules in the presence of a strong magnetic field gradient (Aceto et al., 1970). The magnetomechanical movement of dissolved oxygen in an aqueous medium has been demonstrated experimentally (Lyu et al., 1978; Ueno & Harada, 1982), but, as yet, there are no clear tests of the potential biological consequences of this effect.

(c) As the lamellar phospholipids of cell membranes are diamagnetically anisotropic, the orientational effect of an applied magnetic field exceeding approximately 0.1 - 1 T could significantly perturb membrane transport properties (Labes, 1966). In support of this proposal, direct evidence has been obtained for magnetic field effects on the diffusional properties of liquid crystals (Teucher et al., 1971; Hakemi & Labes, 1974, 1975). Using measured values for the anisotropic diamagnetic susceptibility of model phospholipid membranes (Boroske & Helfrich, 1978), it can be estimated from theoretical considerations that the magnetic interaction energy within a typical cell membrane will exceed the Boltzmann thermal energy, kT, in stationary fields greater than approximately 0.5 T (Tenforde, 1985b). At sufficiently high magnetic field intensities, a perturbation of membrane properties might therefore be expected to occur, with possible consequences for other cellular functions.

(d) The sensitivity of cell membranes to magnetic field interactions may be greatest at phase transition temperatures (Amer, 1965; Aceto et al., 1970). This hypothesis is based on the concept that perturbations introduced by relatively weak magnetic interactions should be amplified near a phase transition temperature at which membrane properties undergo abrupt changes. Some indirect support for this hypothesis was obtained from studies on thermally-induced developmental failure in flour beetles (Amer, 1965), in which higher temperatures were required to elicit developmental wing abnormalities in the presence of a strong magnetic field. More direct evidence for membrane sensitivity to static magnetic fields at phase transition temperatures has recently been obtained (Liburdy et al., 1986; Liburdy & Tenforde, 1986). These investigators observed changes in the permeability of liposome bilayer membranes composed of saturated phospholipids, when the liposomes were exposed to static magnetic fields at temperatures in the prephase transition region from 40 to 40.7 °C. At temperatures of lower than 40 °C or higher than 40.7 °C, no effects on liposome membrane transport were observed in fields as high as 7.5 T.

5.3 Effects on Organs and Tissues

Examples of mammalian tissue and organ alterations that have been observed following magnetic field exposure include changes in:

(a) blood and bone marrow cellular composition (Barnothy et al., 1956; Barnothy & Sumegi, 1969a,b; Nakagawa et al. 1980; Gorchonskaya, 1984);

(b) serum chemistry (Nakagawa et al. 1980; Tvildiani et al. 1983);

(c) microcirculation (Demetsky et al., 1979; Grohmann et al., 1986)

(d) thrombocyte coagulation (Rusyayev, 1979);

(e) electrolyte balance in blood, urine, and various tissues (Hanneman, 1969; Markuze et al., 1973; Tvildiani et al., 1981);

(f) functional and structural properties of various organs and tissues (Reno & Nutini, 1963, 1964; Toroptsev, 1968; Galaktionova & Strzhizhovsky, 1973; Bucking et al., 1974; Wordsworth, 1974; Kholodov & Shishlo, 1980; Strzhizhovsky et al., 1980; Rabinovitch et al., 1983; Strzhizhovsky & Mastryukova, 1983);

(g) immune response (Pautrizel et al., 1969; Kandil & Elashmawy, 1981); and

(h) endocrine regulation (Klimovskaya & Maslova, 1981, 1983; Friedman & Carey, 1972).

With the exception of one study on endocrine changes (Klimovskaya & Maslova, 1983), all of the reported alterations in tissue and organ properties were observed at static magnetic field levels below 1 T. These observations are therefore difficult to reconcile with the growing body of evidence that the development, growth, and homeostatic regulation of mammals is not significantly affected by prolonged exposure to fields of this magnitude.

Many of the experimental reports have been based on studies with small numbers of exposed and control subjects, and often no attempt has been made by the investigators to replicate their experimental results. Furthermore, the magnetic field exposure conditions frequently have not been well documented. In some studies, inadequate controls were used, such as the use of cage-control animals, instead of sham-exposed controls. In this case, effects attributed to magnetic fields may have occurred in response to stresses imposed by other factors, such as adaptation to new caging conditions, differences in ambient temperature, sound levels, lighting conditions, and so forth. There have been few attempts to verify the findings of tissue and organ effects through independent replication in other laboratories. In the few cases where such attempts have been made, the original results have not been successfully replicated. For example, the early reports (Barnothy et al., 1956; Barnothy & Sumegi, 1969a,b; Nakagawa et al., 1980) of haematopoietic alterations have not been confirmed in other studies (Eiselein et al., 1961; Nahas et al., 1975; Viktova et al., 1976; Battocletti et al., 1981). Similarly, earlier reports (Pautrizel et al., 1969; Kandil & Elashmawy, 1981) that magnetic fields alter the immune status of exposed subjects have not been confirmed by Bellossi (1983) or in more recent studies designed to test humoral and cell-mediated immunity in mice exposed for 6 days to a 1.5-T stationary magnetic field (Tenforde & Shifrine, 1984).

It should be added that, where effects have been identified, efforts have seldom been made to explore the consequences for other related end-points, as a means of verifying the previous findings. Furthermore, only a few studies have addressed magnetic field effects on tissues and organs using classical anatomical and physiological methods. In view of these considerations, the existence of deleterious effects of static magnetic fields on tissue and organ functions must, at present, be considered as questionable.

5.4 Effects on the Circulatory System

The occurrence of magnetically induced potentials associated with blood flow in the aorta (Fig. 4, section 4.1) has been demonstrated on electrocardiogram (ECG) recordings from rats (Gaffey & Tenforde, 1981), rabbits (Togawa et al., 1967), dogs and baboons (Gaffey & Tenforde 1979; Gaffey et al., 1980), and monkeys (Beischer & Knepton, 1964; Beischer, 1969; Tenforde et al., 1983) exposed to static magnetic fields (Fig.7). The primary change in the ECG in the field is an alteration of the signal amplitude at the locus of the T-wave. The repolarization of ventricular heart muscle, which gives rise to the T-wave, occurs in the normal ECG at approximately the same time in the cardiac cycle as the ejection of blood into the aorta. It is therefore reasonable to expect that the flow potential induced by the magnetic field is superimposed on the T-wave.

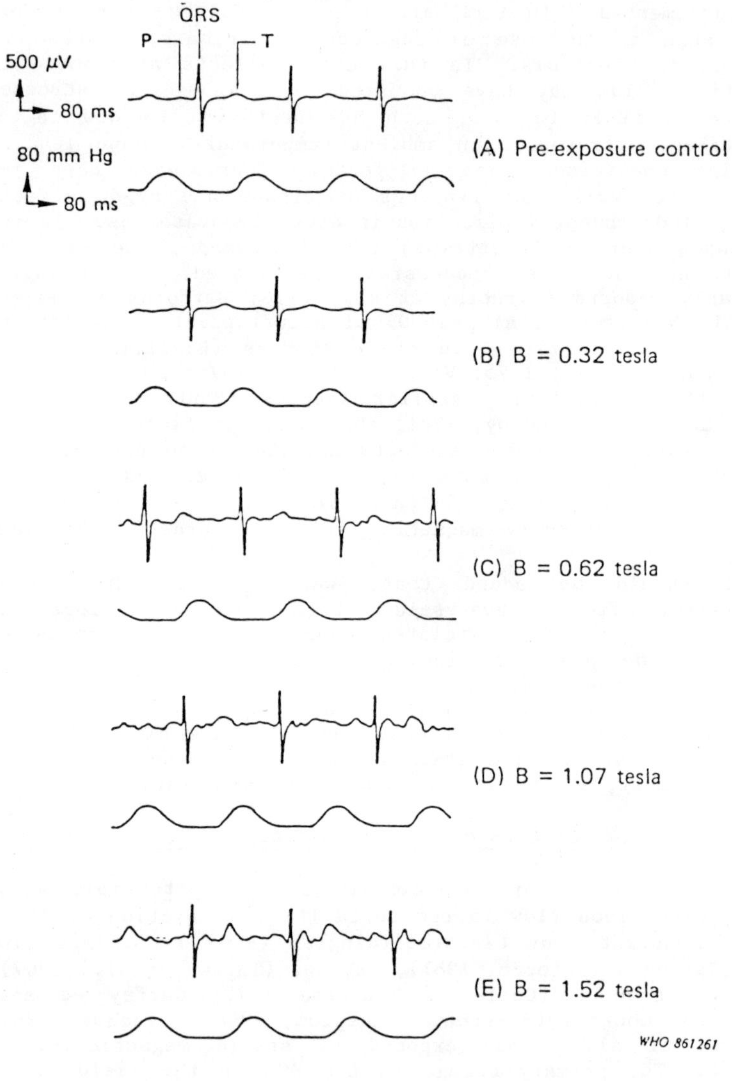

Fig. 7. Electrocardiogram and arterial blood pressure of a monkey exposed to static magnetic fields (B) up to 1.5 T. An increase in T-wave amplitude (Fig. 4) is observed, but no changes in the arterial pressure are demonstrable. From: Tenforde & Budinger (1986).

From the theoretical discussion in section 4, four predictions can be made regarding the induced blood flow potential and the associated magnetohydrodynamic effects:

(a) an induced flow potential should have a linear dependence on the applied magnetic field strength;

(b) the magnitude of the potential should be a function of the orientation of the animal relative to the field direction;

(c) the induced potentials observed in the ECG should increase with the size of the animal;

(d) the resultant magnetohydrodynamic effects should be small.

In the following section, experimental data will be described that relate to these predictions.

5.4.1 Linear relationship of induced flow potential and magnetic field strength

Experimental tests of the linear relationship between the magnetically-induced aortic blood flow potential and the applied magnetic field strength have been carried out by recording the ECG of several species of mammals during exposure to graded field intensities. From the ECG records of rats exposed to static fields ranging from 0.1 to 2.1 T, a field-strength-dependent increase in T-wave amplitude was observed at field levels greater than 0.3 T (Gaffey & Tenforde, 1981). The T-wave signal increase was a linear function of the applied field up to 1.4 T. In dogs (Gaffey & Tenforde, 1979), baboons (Gaffey et al., 1980), and monkeys (Tenforde et al., 1983), the threshold for detection of the T-wave amplitude change was 0.1 T, and the increase in signal strength was a linear function of the magnetic field up to 1 T (Fig. 7).

These data support the concept that the T-wave alteration is a consequence of the superposition of an induced aortic blood flow potential, which is theoretically predicted to have a strictly linear dependence on the magnetic field strength.

5.4.2 Induced flow potentials and field orientation

From theoretical considerations, it is predicted that the magnitude and the sign of the induced flow potential should be a function of the angle between the direction of blood flow and the direction of the applied magnetic field. Consistent with this prediction, it has been shown for rabbits (Togawa et al., 1967) and for rats (Gaffey & Tenforde, 1981) that the amplitude of the T-wave signal can be increased, decreased, or remain unchanged by the superimposed aortic blood flow potential, depending on the orientation of the animal relative to the field. It has also been demonstrated that the maximum change in the T-wave amplitude occurs when the long axis of a rat, and hence its ascending aortic vessel, is oriented perpendicular to the field (Gaffey & Tenforde, 1981). This observation is completely consistent with the theoretical prediction that the magnitude of the magnetically-induced aortic blood flow potential should achieve its maximum value when the flow vector and the magnetic field vector are orthogonal.

5.4.3 Dependence of induced blood flow potentials on animal size

The theoretical considerations in section 4 suggest that the magnitude of induced aortic blood flow potentials should be significantly greater for large animal species than for the rodent. When ECG measurements were made on animals exposed to a 1-T field, with an orientation perpendicular to the body axis, the maximum aortic flow potentials recorded at the body surface were 75 μV for rats (average weight 0.25 kg) (Gaffey & Tenforde, 1981), 175 μV for baboons (5 kg) (Gaffey et al., 1980), 200 μV for monkeys (5 kg) (Tenforde et al., 1983), and 390 μV for dogs (9 kg) (Gaffey & Tenforde, 1979). Thus, greater magnetically-induced blood flow potentials were observed with larger animal species, conforming to theoretical expectations.

5.4.4 Magnetohydrodynamic effects

A test of potential alterations in haemodynamic parameters as a consequence of magnetohydrodynamic interactions was made by recording the arterial blood pressure of monkeys during exposure to homogeneous, static magnetic fields ranging from 0.1 to 1.5 T. The study was conducted with an accuracy of ± 2 mmHg in the recording of systolic and diastolic blood pressures. No measurable alteration in blood pressure was observed in fields up to 1.5 T (Fig. 7). This observation is consistent with the theoretical prediction that negligible haemodynamic alterations result from magnetohydrodynamic interactions with blood flow in fields of less than 2 T (Tenforde et al., 1983).

5.4.5 Cardiac performance

Several indices of cardiac function have been studied in order to assess the possible physiological effects of the electrical potentials induced by an applied magnetic field. These indices include blood pressure, heart rate, and the bioelectric activity of heart muscle. As described above, there is no measurable alteration in the blood pressure of monkeys exposed to a 1.5-T stationary field. The heart rate and electrical properties of heart muscles have been determined from ECG measurements on rats exposed to stationary fields up to 2.1 T (Gaffey & Tenforde, 1981), rabbits in a 1-T field (Togawa et al., 1967), dogs (Gaffey & Tenforde, 1979) and baboons (Gaffey et al., 1980) in fields up to 1.5 T, and monkeys exposed to fields of up to 1.5 T (Tenforde et al., 1983) and to a 10-T field (Beischer, 1969). Significant changes in heart rate were not observed during acute magnetic field exposures in any of these studies. Similarly, the amplitudes of the P, Q, R, and S waves of the ECG were not altered, indicating that the applied magnetic field had no effect on the depolarization characteristics of the auricular and ventricular heart muscle. The data from these studies on various species of animals also indicated that no cardiac arrhythmias occurred during acute exposures to the field levels studied.

These experimental observations provide evidence that little or no cardiovascular stress should result from exposure to the highest static magnetic field levels routinely encountered by man. However, this conclusion must be tempered by the recognition that no data are available in the literature relating to cardiovascular performance during protracted exposure to large stationary magnetic fields. Also, from the theoretical considerations discussed in section 4, it would be anticipated that measurable haemodynamic pertubations could occur during exposure to static fields that significantly exceed 2 T. For example, it has been predicted theoretically (Tenforde, 1985a) that a 5-T field would produce a reduction in aortic blood flow velocity of up to 7% in a human adult.

5.5 Nervous System and Behaviour

On the basis of the theoretical models described in section 4, it is not anticipated that stationary magnetic fields with intensities up to 2 T would produce measurable alterations in nerve bioelectric properties. The theoretical expectations agree with the existing experimental information on the behaviour of isolated neurons in large static magnetic fields.

5.5.1 Excitation threshold of isolated neurons

From the theoretical considerations of Wikswo & Barach (1980), it can be estimated that a static magnetic field of at least 24 T would be needed to reduce the velocity of action potential conduction in isolated neurons by 10%. The threshold for neural excitation has been examined for both intact frog sciatic nerves and single myelinated sciatic nerve fibres during exposure to a homogeneous, static magnetic field (Liberman et al., 1959; Gaffey & Tenforde, 1983). In both studies, the field orientation was transverse to the nerve axis. No evidence was obtained in these studies of an effect of a 1-T magnetic field on the minimum electrical stimulus required to evoke action potentials in either single fibres or intact sciatic nerves.

An important observation that has a direct bearing on such studies was made by Gaffey & Tenforde (1983), who determined the temperature coefficient of the frog sciatic nerve excitation threshold, and found it to rise with increasing temperature. To obtain reliable results, it was found that the temperature must be controlled to within 0.1 °C.

5.5.2 Action potential amplitude and conduction velocity in isolated neurons

Several groups of investigators have studied the properties of evoked action potentials in isolated nerve preparations during exposure to static magnetic fields oriented either parallel or perpendicular to the nerve axis. Schwartz (1978) exposed the circumoesophageal nerve of the lobster to static fields with a maximum strength of 1.2 T. The nerve preparation was maintained in an L-shaped chamber, and the field gradient along the sections of nerve oriented parallel and perpendicular to the field lines were 2 and 15 T/m, respectively. No effects of either the parallel or perpendicular fields, applied for periods of up to 30 min, were observed on the nerve conduction velocity. Gaffey & Tenforde (1983) conducted similar measurements on intact sciatic nerves exposed to either parallel or perpendicular 2-T static fields that were homogeneous to within 0.1% over the entire length of the nerve. With both field configurations, no effects were observed during a continuous 4-h exposure on either the amplitude or the conduction velocity of maximal evoked action potentials. Extending the duration of exposure to 17 h was also found not to influence on the impulse conduction velocity.

Schwartz (1979) measured the membrane potentials and transmembrane currents in lobster circumoesophageal nerves exposed to a 1.2-T stationary field. Both parallel and perpendicular field orientations relative to the nerve axis were used in these experiments, and the field gradients were identical to those described above in the discussion of Schwartz's studies on nerve conduction velocity (Schwartz, 1978). No effects of the parallel or perpendicular magnetic fields were observed on either the action potentials or the transmembrane currents during nerve excitation.

In contrast to the negative results of the studies described above, the results of two other studies have shown effects of static magnetic fields on nerve bioelectric activity (Reno, 1969; Edelman et al., 1979). However, Tenforde (1985b) suggested that the apparent magnetic field effects observed in these studies are probably attributable to a lack of precise temperature control, the importance of which has already been discussed above.

5.5.3 Absolute and relative refractory periods of isolated neurons

Following the passage of a maximal action potential, an isolated peripheral nerve enters an absolute refractory period of 1 - 2 ms duration, during which a second impulse cannot be evoked. Following the absolute refractory period, the nerve enters a relative refractory period during which action potentials of progressively increasing amplitude can be evoked by electrical stimulation. After a period of approximately 4 - 6 ms, the second action potential reaches the same maximal amplitude as the impulse elicited by the initial stimulus, thus marking the end of the relative refractory period. The characteristics of both the absolute and relative refractory periods have been examined during the exposure of frog sciatic nerves to a homogeneous 2-T field (Gaffey & Tenforde, 1983; Tenforde et al., 1985). Using both parallel and perpendicular configurations of the magnetic field relative to the nerve axis, no influence of the field was observed on the duration of either the absolute or the relative refractory periods. In addition, the amplitudes of impulses evoked during the relative refractory period were unaffected by the magnetic field exposure.

In summary, the majority of the experimental studies that have been conducted to date indicate that static magnetic fields up to 2 T have little or no influence on the bioelectric properties of isolated neurons.

5.5.4 Effects of static magnetic fields on the electroencephalogram

Several reports have been made of changes in brain electrical activity during the exposure of experimental animals to static fields ranging from approximately 0.1 to 9.1 T. The information is inconsistent, at times contradictory, and requires additional investigations before a definite judgement can be made.

In a series of electroencephalogram (EEG) studies on squirrel monkeys, Beischer & Knepton (1966) observed that exposure to static magnetic fields produced a significant increase in the amplitude and frequency of brain electrical signals recorded below the scalp in the frontal, parietal, temporal, occipital, and median cranial regions. Recordings of the EEG were made in homogeneous fields with field strengths ranging from 1.47 to 9.13 T. EEG recordings were also made in strong gradient fields. During exposures ranging from 3 to 45 min, it was found that the predominant EEG frequencies shifted from their pre-exposure range of 8 - 12 Hz to 14 - 50 Hz, independently of the field intensity or homogeneity. The amplitude of the signals also increased from the control level of 25 - 50 μV to 50 - 400 μV. These changes were uniformly observed in the different cranial regions, which were simultaneously monitored; there was no latency in the response on application of the field. When the field was removed, both the amplitude and frequency spectrum of the EEG signals returned to their pre-exposure levels, indicating the transient nature of this effect.

In analysing the results of their studies, Beischer & Knepton (1966) considered several potential sources of artifacts, including ripple currents from the magnet power supply, animal movements associated with heart contractions and breathing, pick-up of stray 60-Hz fields by the EEG electrodes and leads, and skeletal muscle tremors. All of these factors, except for muscle tremors, could be excluded because their characteristic frequencies were outside the frequency range observed for the predominant EEG signals in the presence of a static magnetic field. However, the characteristics of the EEG tracings obtained from monkeys in the magnetic field suggest that "myographic noise" from skeletal muscles may have been superimposed on the brain electrical signals. It is also possible that other uncontrolled factors, present only during excitation of the magnet coils, including mechanical vibrations, audible noise, and an increased ambient temperature, could have led to an altered pattern of brain electrical activity.

In contrast to the above findings with monkeys, Kholodov reported that the exposure of rabbits to relatively weak static fields (0.08 - 0.10 T) produced an EEG signature characteristic of a general inhibitory state in the central nervous system (Kholodov, 1964, 1966; Kholodov et al., 1969).

The major changes in the EEG during magnetic field exposure were the occurrence of slow waves and high-amplitude spindles observed in the electrical activity recorded from different regions of the brain. The phenomenon was not uniformly exhibited in all of the tests conducted by Kholodov; in a series of 100 field exposures on 12 rabbits, the author observed the occurrence of spindles in 30% of the tests, and an increase in the number of slow waves with frequencies of less than 4 Hz in 19% of the tests (Kholodov, 1966). The percentage of animals exhibiting EEG responses to the field was not stated. Both spindles and slow waves in the EEG occurred with a latency of approximately 15 s after the field was turned on, and reached maximum levels after 45 s of exposure. The increased number of spindles and slow waves persisted during exposure to a 0.1-T field for 3 min, and decreased immediately after the field was turned off. However, 15 to 25 s after the exposure was terminated, a transient increase in the number of spindles and slow waves occurred with a duration of approximately 20 - 30 s.

Kholodov (1966) presented evidence that the EEG alterations observed in his studies on rabbits were not artifacts resulting from the induced potentials that occur during the switching on and off of an electromagnet. This possibility was excluded on the basis of trials in which the magnet was energized and de-energized at various rates, with no resulting change in the character of the observed EEG alterations.

Kholodov (1966, 1981, 1982) also described histological changes in the brains of mammals exposed to static magnetic fields for brief periods. The significance of these anatomical changes was not clearly established. The differences in the results obtained by Beischer & Knepton (1966) and by Kholodov (1966) may be related to the one order of magnitude difference between the field strengths used in their studies. Furthermore, Battocletti et al. (1981) did not find changes in potentials evoked by stimulation of extremities in rhesus monkeys exposed to 2 T for 48 h.

The positive observations may be explained by additive effects on elements of the central nervous system (Valentinuzzi, 1965). It should be noted that no recent studies on the effects of static magnetic fields on the bioelectric activity of the brain were found in the published literature. It seems that this area deserves further study using modern electroencephalographic methods. The application of recordings from single locations in the brain to elucidate the neural basis for sensitivity to magnetic cues in pigeons (Semm et al., 1984; Semm, 1986) may serve as an example.

5.5.5 Behavioural effects

An inherent sensitivity to the weak geomagnetic field and correlated behavioural responses has been demonstrated for a number of different organisms and animal species. It has been well documented experimentally that weak magnetic fields influence the migratory patterns of birds (Keeton, 1971; Emlen et al., 1976; Bookman, 1977), the kinetic movement of molluscs (Ratner, 1976), the waggle dance of bees (Martin & Lindauer, 1977), the direction-finding of elasmobranch fish (Kalmijn, 1978, 1982), and the orientation and swimming direction of magnetic bacteria (Blakemore, 1975; Blakemore et al., 1980). The mechanisms underlying the magnetic sensitivity of elasmobranchs and magnetotactic bacteria have been discussed in sections 4.1.1 and 4.1.2.2.

A precise mechanism underlying the magnetic sensitivity of other organisms has not been elucidated, although small deposits of magnetite crystals have been discovered in the cranium of pigeons (Walcott et al., 1979), in the tooth denticles of molluscs (Lowenstam, 1962; Kirschvink & Lowenstam, 1979), and in the abdominal region of bees (Gould et al., 1978). Magnetite has been also reported in various anatomical sites in dolphins (Zoeger et al., 1981), tuna (Walker et al., 1984), butterflies (Jones & MacFadden 1982), turtles (Perry et al., 1981), mice (Mather & Baker, 1981), and human beings (Kirschvink, 1981; Baker et al., 1983). The possible role of magnetite in the geomagnetic direction-finding mechanism possessed by some of these species has not been established, nor is it clear for all of the mammalian species in which magnetite deposits have been reported to occur (Baker, 1980; Gould & Able, 1981).

Although the directional cues derived from the weak geomagnetic field by certain species of animals have been demonstrated by careful study, the possible effects of magnetic fields on the behaviour of higher organisms are by no means established. Several studies with rodents have reported effects of static magnetic fields of less than 1 T on locomotor activity and patterns of food and water consumption (Aminev et al., 1967; Russell & Hendrick, 1969; Pelyhe et al., 1973; Nakagawa, 1979; Nakagawa et al., 1980; Shust et al., 1980). In contrast to these earlier reports, Davis et al. (1984) did not observe any behavioural abnormalities in mice exposed for prolonged periods to a 1.5-T field. The behavioural end-points examined in this extensive study included memory retention of an electroshock-motivated passive avoidance task, general locomotor activity, and sensitivity of the subjects to a neuropharmacological agent (pentylenetetrazole). Smirnova (1982) also did not find any behavioural effects in rats exposed to 0.3 T or 1.6 T for 5 min/day for 3 successive days.

The effects on primate behaviour of exposure to intense magnetic fields was studied by Thach (1968). In one study, 3 squirrel monkeys (Saimiri sciureus) were conditioned to respond to a visual vigilance task and subsequently exposed to static magnetic fields in the core of a water-cooled Bitter magnet. Response was greatly suppressed by fields of 7 T or more. A threshold seemed to exist between 4.6 and 7 T. In a second study (deLorge, 1979), 8 squirrel monkeys were trained in several operant tasks and a similar suppression response was observed in fields up to 9.7 T. In addition, 2 of the monkeys regurgitated when exposed to these higher fields. All of these effects were reported to be reproducible.

5.6 Visual System

As discussed in section 4.2, one of the most clearly established magnetic field effects in biological systems is the phenomenon of magnetophosphenes, in which a flickering light is produced in the visual field during exposure to time-varying magnetic fields.

Although the phenomenon of phosphenes has not been reported by human observers during exposure to large static magnetic fields, there are two potential interaction mechanisms between these fields and elements of the retina that are involved in the visual response to photic stimulation. First, the photoreceptor outer segments are subject to orientation in a static magnetic field as the result of their large diamagnetic anisotropy (Chalazonitis et al., 1970; Hong et al., 1971; Becker et al., 1978b; Hong, 1980). Second, the initial photoisomerization event elicited by photon absorption in retinal photopigments is followed by a series of ionic fluxes that lead to excitation of the retinal neurons, and ultimately the visual cortex via a complex neural pathway. This component of the phototransduction process could be influenced by static magnetic fields as the result of ionic current distortion and/or inductive effects, as discussed in section 4. However, electrophysiological studies on the retinal response to photic stimuli in cats and monkeys have not shown any effects of exposure to a 1.5-T static magnetic field (Gaffey & Tenforde, 1984; Tenforde et al., 1985).

That photoreceptors may play a crucial role in magnetoreception is suggested by the fact that inhibitory effects of low magnetic fields on pineal melatonin synthesis were not found in albino rats (Olcese et al., 1985) or in rats exposed to magnetic fields in total darkness (Reuss & Olcese, 1986), thus supporting the theory of Leask (1978) that incident radiation is an important factor in magnetic field sensitivity, i.e., light might be essential to the process of magnetoreception. In contrast to rats (both albino and pigmented strains), hamsters did not respond to magnetic stimuli, as measured by the inhibition of pineal gland metabolism (Olcese & Reuss, 1986), and species-specific differences as well as reciprocal effects between photoreceptors and retinal pigments should be taken into consideration.

Since birds possess a direction-finding sense that appears to be based on simultaneous detection of the earth's magnetic and gravitational fields, Semm et al. (1984) undertook a study to explore possible neural mechanisms for the integration of magnetic and gravitational cues. Leask (1977) proposed that the magnetic field compass was located in the retina of the bird. Thus, Semm et al. (1984) recorded single unit electrical activity in the lateral and superior vestibular nuclei, the vestibulo-cerebellum, and the nucleus of the basal optic root, which has a projection to the vestibular system, in pigeons, under magnetic stimulation by fields of about 42 μT. The responses of these cell systems were direction-selective, i.e., different cells responded to different directional changes in the magnetic field. The interpretation of this was that magnetic cues may be conveyed from the visual to the vestibular system via a projection from the basal optical root, and then related to the movement of the bird.

The effects of static magnetic fields on turtle retinas *in vitro* were studied by Raybourn (1983) (see also Tenforde *et al.*, 1985). No changes were seen in electroretinograms (ERG) from dark- or light-adapted eyes during exposure to 1-T fields. However, 2- to 3-mT fields suppressed the B-wave of the ERG in eyes prepared during the light-to-dark adaptation phase, which lasts for about 2 h. No effects of 1.5-T fields on the ERG in cats and monkeys were observed (Gaffey & Tenforde, 1984; Tenforde et al., 1985), but circadian variations were not studied. These findings have not been interpreted. The static magnetic field strength at which effects were noted in turtle retinas was too low to influence ionic fluxes that occurred in the retina following stimulation by light.

5.7 Physiological Regulation and Circadian Rhythms

In assessing the responses of living organisms to static magnetic fields, an important aspect is the maintenance of normal homeostatic regulation. The literature on this subject is often contradictory. For example, the finding by Sperber et al. (1984) that thermoregulation in rodents is affected by strong magnetic field spatial gradients, could not be replicated by Tenforde (1986c). It should be noted that Gremmel et al. (1984) described changes in thermoregulation in human beings exposed to magnetic fields. One of the central issues in this assessment is whether exposure to magnetic fields produces an alteration in the normal circadian rhythm of major physiological and behavioural variables. Several of the investigations discussed in this section indicate that exposure of mammals to static magnetic fields may lead to hormonal alterations and to other metabolic effects that could potentially affect physiological regulation, and thereby lead to an alteration in the normal circadian rhythm. Although there is relatively little information available on this subject, several reports in the literature suggest that weak magnetic fields may influence circadian regulation.

Brown & Scow (1978) observed a modulation of the normal 24-h circadian activity period in hamsters, when a weak magnetic field with a maximum intensity of 26 μT was applied in 26-h cycles. The nocturnal sensitivity of mice to morphine was found by Kavaliers et al. (1984) to be diminished, when the animals were exposed to a rotating magnetic field with an intensity ranging from 105 μT to 9 mT. A cancellation of the earth's magnetic field by Helmholtz coils was found to alter the circadian activity of birds (Bliss & Hepner, 1976). It has recently been reported that artificial changes in the strength and direction of the local geomagnetic field are sufficient to alter the electrical activity of pineal cells in the guinea-pig (Semm et al., 1980; Semm, 1983), rat (Reuss et al., 1983), and pigeon (Semm et al., 1982, 1984; Semm, 1983, 1986). In related studies, it was demonstrated in albino rats that artificial changes in the ambient magnetic field reduced the nocturnal rise in pineal melatonin contents and the activity of the involved enzymes, N-acetyltransferase (Welker et al., 1983; Olcese et al., 1985; Olcese & Reuss, 1986) and hydroxyindole-O-methyltransferase (Reuss & Olcese, 1986). Interestingly, this effect was not found using NMR-strength fields of 0.14 T (Reuss et al., 1985).

In other recent studies (Tenforde, 1985c; Tenforde et al., 1986b,), prolonged exposures of mice to a 1.5-T static magnetic field did not produce any alterations in the circadian rhythm of several physiological and behavioural variables. Noninvasive transducer techniques were used to provide continuous measurements of core body temperature, respiration, body mass, food intake and excreta, and two independent indices of locomotor activity. The rodents were subjected to a homogeneous 1.5-T field under 3 different exposure regimens: (a) continuous exposure for 5 days; (b) intermittent exposure in an 8 h-on/16 h-off cycle for 10 consecutive days; and (c) serial exposures to the field under the 5-day continuous and 10-day intermittent schedules. In addition, the sensitivity of circadian oscillations to a 1.5-T field was tested both in mice that were maintained on a diurnal light/dark cycle, and in mice that were placed in a free-running circadian state by the maintenance of continuous dim illumination. Under all of these conditions, no influence of a 1.5-T field was observed on the circadian variations in any of the physiological or behavioural parameters studied.

In an effort to elucidate whether static magnetic fields perturb the light-elicited electrical activity of the retina, Raybourn (1983) recorded the external ERG of isolated turtle retinas during light stimulation in the presence of magnetic fields of graded strength. When the retinal preparations from light-adapted or dark-adapted eyes were studied, no changes in the ERG occurred in fields up to 1 T. However, the amplitude of the ERG b-wave, which results from the electrical activity of nerve cells in the retina, was consistently suppressed in retinas prepared during the light-to-dark transition phase of the diurnal 12 h-light/12 h-dark cycle. During this transition phase, which extended for approximately 2 h after the onset of darkness, the photoreceptor cells underwent rapid changes in both physiological and metabolic activities (Bubenik et al., 1978; Young, 1978).

The magnetic field effect was observed with intensities as low as 2 - 3 mT, and was rapidly reversible following termination of exposure. This effect was observed in both the cone-dominant retinas of *Pseudemys scripts* turtles, and the mixed rod-cone retinas of *Chelydra serpentina* turtles, suggesting that it is independent of the photoreceptor cell type. The circadian dependence of the magnetic field sensitivity was clearly demonstrated by studies in which the light/dark cycle was phase shifted by several hours (Tenforde et al., 1985).

An alteration in human twilight visual acuity has been reported to occur in response to changes in the strength of the ambient geomagnetic field (Krause et al., 1984). It has been suggested that this visual alteration may have its origin in a quantum mechanical effect on biochemical reactions in the retina, similar to that discussed by Schulten et al. (1978).

5.8 Genetics, Reproduction, and Development

Developing organisms frequently exhibit a strong response to noxious environmental factors. This observation has stimulated a relatively large number of studies on the potential effects of static magnetic fields on the genetics, reproduction, and development of various organisms. Investigations on a variety of non-mammalian test systems have led to several reports of mutagenic and developmental effects resulting from exposure to both gradient and homogeneous magnetic fields. Effects observed with strong magnetic field gradients have included alterations in the sex ratio and development of *Drosophila* pupae (Mulay & Mulay, 1964; Markuze et al., 1973; Tvildiani et al., 1981), and abnormal development of sea urchin, frog, and salamander eggs (Perakis, 1947; Neurath, 1968; Levengood, 1969; Ueno et al., 1984). Inhibition of limb regeneration in crabs (Lee & Weis, 1980) has also been observed. Homogeneous magnetic fields have been reported to alter the development of chicken embryos (Joshi et al., 1978), and guppies (Brewer, 1979), and the rate of fertilization of trout eggs (Strand et al., 1983). It is interesting to note that Perakis (1947) did not find any effects of a homogeneous 3.3-T field on the development of sea urchin eggs, and Ueno et al. (1984) did not observe any effects of a 1-T homogeneous field on the development of frog embryos. The absence of effects of homogeneous magnetic fields on frog egg development is also supported by the experimental observations of Iwasaki et al. (1978) and Mild et al. (1981). In contrast, developmental abnormalities were observed in both sea urchin eggs and frog embryos exposed to large magnetic field gradients (Perakis, 1947; Ueno et al., 1984). Ueno et al. (1984) suggested that the developmental effects of gradient fields may result from a redistribution of dissolved oxygen or from the orientation of mitochondrial cytochromes in large magnetic fields with gradients exceeding 10^4 T/m.

In studies on mammals, it has been reported that homogeneous and gradient fields up to 0.94 T inhibit weight gain in young mice and produce weight loss in older animals (Barnothy, J.M., 1964). The rate and number of live births and the average birth weight have also been reported to decrease following prenatal and postnatal exposure of mice to a homogeneous field (Nakagawa, 1979). In contrast to these reports, studies on young mice exposed for up to 15 days to a nearly homogeneous field with a maximum strength of 1.44 T did not reveal any effects on growth rate (Eiselein et al., 1961). Bellossi et al. (1984) did not observe any variations in growth of either mice or rats exposed to static magnetic fields of up to 800 mT for up to 250 days. The intrauterine exposure of mice and rats to either a 1-T homogeneous field or a 2.5-T/m gradient field was also found not to influence fetal or postnatal development (Sikov et al., 1979).

Exposure of mice to static magnetic fields of 1.6 T, during a 30-day period, resulted in reversible changes in spermatogenic epithelium and in a considerable decrease in the number of mature germ cells (Galaktionova et al., 1985). These and other authors (Toroptsev et al., 1974; Udintsev & Khlynin, 1979) considered the testes a vulnerable organ when exposed to static or time-varying (20 mT, 50 Hz) magnetic fields. Morphological changes in the testes and other organs, which occur after a 6-h exposure to magnetic fields, revert to normal after approximately one month.

Several studies have been carried out to determine whether genetic defects can be detected following magnetic field exposure. No increase in mutation frequency was observed by Kale & Baum (1979) among the progeny of Drosophila males exposed as eggs, larvae, pupae, and adults to 1.3 - 3.7-T homogeneous magnetic fields. Similar results were obtained by Mittler (1971) and Diebolt (1978), who exposed Drosophila males to fields of 1 - 1.1 T. Baum et al. (1979) also found that exposure of the plant Tradescantia to homogeneous fields up to 3.7 T did not lead to any increase compared with controls in three mutagenic indices, namely, pollen abortion, micronuclei formation, and pink stamen hair production. Dominant lethal assays have been conducted by Mahlum et al. (1979) with male mice exposed to either a uniform 1-T or a 2.5-T/m gradient field for 28 days prior to mating. No effects of exposure to either the homogeneous or the gradient field were observed. This result and the study of Strzhizhovsky et al. (1980) indicate that such exposure does not induce chromosomal aberrations in male germ cells.

Recent studies have also demonstrated that the exposure of cultured Chinese hamster ovary cells to a 0.35-T homogeneous field does not lead to alteration in DNA synthesis or chromosome structure (Wolff et al., 1980). The structure and biological activity of bacteriophage DNA have also been found to be unaffected by exposure to a 2-T homogeneous field (Roots et al., 1982).

5.9 Conclusions

Studies on the effects of static magnetic fields on enzyme reactions and cellular and tissue functions have provided diverse, and often contradictory, findings. Nevertheless, available evidence indicates that there are few irreversible effects on such systems, with the possible exception of:

(a) enzymes and photosynthetic systems that involve radical-mediated reaction intermediates; and

(b) cellular systems in which the membrane is undergoing a structural phase transition during magnetic field exposure.

The occurrence of significant genetic or developmental alterations in cellular tissues and animal systems exposed to high-intensity static magnetic fields appears unlikely from available evidence. One possible exception relates to unconfirmed reports of alterations in the embryonic development of amphibian species exposed to strong magnetic field gradients.

The magnetic induction of electrical potentials and currents in the central circulatory system does not produce measurable cardiovascular stress during short-term exposure to static fields of up to 2 T. This conclusion must be tempered by the recognition that data do not exist on cardiovascular performance during protracted magnetic field exposure.

The majority of experimental studies conducted with isolated neurons indicate that static magnetic fields of up to 2 T have no irreversible influence on neuronal bioelectric properties. Several reports have referred to changes in brain electrical activity and behaviour in animals exposed to fields ranging from 0.1 to 9 T, but the data are inconsistent and at times contradictory.

An inherent sensitivity to the weak geomagnetic field and correlated behavioural responses have been demonstrated for a number of different organisms and animal species. However, behavioural effects in higher organisms have not been established at field strengths of less than 2 T. Although the data are inconsistent, effects on physiological regulation and circadian rhythms have been reported in animals, due to alterations in the local geomagnetic field. Negative findings in higher organisms have been reported in studies involving field levels as high as 1.5 T.

Thus, reversible or transient effects have been reported in lower animals due to exposure to low-intensity static fields or due to alterations in the ambient geomagnetic field. However, no irreversible effects have been established due to static magnetic field exposures of up to 2 T.

6. BIOLOGICAL EFFECTS OF TIME-VARYING MAGNETIC FIELDS

An extensive literature exists on the response of animals and isolated cellular and tissue systems to ELF magnetic fields. At present, this body of research is difficult to interpret in a systematic manner because of two factors:

(a) A wide range of intensities, frequencies, waveforms, and exposure durations have been used. Many of the earlier studies involved sinusoidal fields oscillating at frequencies below 100 Hz, but research during the last several years has focused increasingly on the biological effects of square-wave or pulsed fields with complex waveforms.

(b) Very few of the reported effects of ELF magnetic fields have been independently replicated in different laboratories.

In spite of these difficulties, there is a growing body of evidence that suggests that living systems exhibit a response to ELF magnetic fields under conditions in which the field intensity and rate of change in time (dB/dt) are sufficient to induce currents greater than the naturally occurring levels in tissues and extracellular fluids. This effect is best illustrated by the phenomenon of magnetophosphenes, which is the one well established biological effect of ELF magnetic fields. Although less well established, there is also evidence suggesting that pulsed magnetic fields, such as those used clinically to facilitate bone fracture reunion, may exert direct biological effects through the induction of tissue currents that exceed the endogenous levels.

The following topics are summarized in this section:

(a) magnetophosphene research;

(b) studies on the nervous system and animal behaviour;

(c) cellular, tissue, and animal responses to magnetic fields with various waveforms and repetition frequencies in the ELF range;

(d) studies on the effects of pulsed magnetic fields on bone growth and repair; and

(e) thresholds for biological effects as a function of field frequency and induced current densities.

6.1 Visual System

High-intensity magnetic fields oscillating in the ELF range produce visual sensations in human subjects that are known as magnetophosphenes. This phenomenon has already been discussed in section 4.2.

Studies (Silny, 1981, 1984, 1986) have been performed to characterize ELF magnetic-field effects on visually evoked potentials (VEP). Fields greater than 50 mT in the frequency range below 100 Hz were demonstrated to reverse the polarity and reduce the amplitude of VEP recorded from human volunteers. This effect persisted after the termination of the magnetic field exposure. Approximately 40 min after the magnetic field exposure, the VEP was found to return to a normal form. This effect of magnetic fields was shown to be frequency-dependent, the field strength required to elicit an alteration in the VEP decreasing as the field frequency increased from 5 to 100 Hz. It should be noted that the field strength required to alter the VEP is approximately one order of magnitude greater than that required to elicit clear magnetophosphene patterns.

6.2 Studies on Nerve and Muscle Tissue

Several studies have been made on the electrical response of neurons to stimulation with time-varying magnetic fields. As discussed by Bernhardt (1979, 1985), the current densities induced by the field must exceed 1 - 10 mA/m² in order to have an appreciable effect on the nerve bioelectric activity, and a threshold extracellular current density of about 20 mA/m² has been found experimentally with Aplysia pacemaker neurons stimulated by an ELF electric field (Wachtel, 1979). In a subsequent study with Aplysia (Sheppard, 1983), an induced current density of approximately 5 mA/m² produced by a 10-mT, 60-Hz sinusoidal field was ineffective in altering the spontaneous neuronal electrical activity. Ueno et al. (1981) were also unable to alter the amplitude, conduction velocity, or refractory period of evoked action potentials in lobster giant axons, by applying sinusoidal magnetic fields with intensities of 1.2 T at 5 - 20 Hz, 0.8 T at 50 Hz, and 0.5 T at 100 Hz.

Using magnetic flux densities in the range of 0.2 - 0.8 T, Kolin et al. (1959) were able to stimulate frog nerve-muscle preparations at field frequencies of 60 and 100 Hz. Oberg (1973) and Ueno et al. (1978) were also able to stimulate contractions in frog nerve-muscle preparations by using pulsed magnetic fields with pulse durations of less than 1 ms. In addition, the excitation of frog sartorius and cardiac muscles (Irwin et al., 1970) and of the sciatic nerves of dogs and rabbits (Maass & Asa, 1970) has been reported to occur in response to pulsed magnetic fields. On the basis of electromyographic recordings from the human arm, Polson et al. (1982) were able to characterize the pulsed magnetic field parameters that elicited a neural response. They indicated that the threshold rate of change of the magnetic field (dB/dt), which was necessary to stimulate the major nerve trunks of the arm, was approximately 10^4 T/s. These fields were discrete pulses, 180 µs long, which will result in a high threshold compared with that for a continuously applied sinusoidal stimulus of between 10 and 100 Hz.

A threshold of perception of about 2×10^3 T/s was reported (McRobbie & Foster, 1984) in human volunteers whose forearms were exposed to a damped sinusoidal magnetic field (2 - 3 cycles of a period equal to 0.3 ms). The currents induced in the peripheral tissues of the forearm were calculated to be approximately 5 A/m^2.

Other effects of time-varying magnetic fields on electrically excitable tissue have been summarized by Bernhardt (1985) (section 4.2). The frequency dependence of these effects has been described by Bernhardt (1985, 1986) (section 8.2).

From these studies, it appears that sinusoidal magnetic fields with intensities in the range generally used in the laboratory or well above the levels encountered by human beings in occupational settings or in the home environment, are insufficient to alter the bioelectric properties of isolated neurons. However, direct magnetic stimulation of nerve and muscle tissues can be achieved by using pulsed fields with a rapid time rate of change of the magnetic flux density. It should also be borne in mind that the effects of sinusoidal fields on complex, integrated neuronal networks, such as those within the central nervous system, may be considerably greater than the effects that occur in single neurons or nerve bundles. This amplification of a field effect could occur through a summation of the small responses evoked in individual neuronal elements (Valentinuzzi, 1965). An additive response mechanism may also underlie the production of magnetophosphenes through the stimulation of multiple neuronal elements of the retina (Valentinuzzi, 1962).

6.3 Animal Behaviour

During the past two decades, a large number of studies on animal behavioural responses to ELF magnetic fields have been reported. A chronological listing of these reports and a summary of the principal findings are given in Table 9.

Several studies in which the behaviour of honeybees and birds was observed to be altered in the presence of combined

Table 9. Behavioural effects of exposure to time-varying magnetic fields

Reference	Subject	Exposure conditions[a]	Results
Friedman et al. (1967)	human being	0.1 and 0.2 Hz, 0.5 - 1.1 mT; acute exposures	Increased reaction time in 0.2-Hz field
Caldwell & Russo (1968)	honey bee	60 Hz, 2.2 - 30 mT; 10-min exposures	Altered exploratory behaviour
Persinger (1969)	rat	0.5 Hz, 0.3 - 3 mT; rotating field; exposure during entire gestational period	Decreased open-field activity and increased defecation when tested postnatally at 21 - 25 days
Persinger & Foster (1970)	rat	0.5 Hz, 0.3 - 3 mT; rotating field; exposure during entire gestational period	Decreased avoidance of aversive electrical shock when tested postnatally at 30 days
Grissett & deLorge (1971)	monkey	45 and 75 Hz, 0.3 mT; fields applied in 10 daily sessions of 1 h duration	No effect on reaction time
Grissett (1971)	monkey	45 Hz, 1 mT; continuous exposure for 42 days	No effect on reaction time
Persinger & Pear (1972)	rat	0.5 Hz, 0.3 - 3 mT, rotating field; exposure during entire gestational period	Suppressed rate of response to a conditioned stimulus preceding an aversive shock when tested postnatally at 70 days
Persinger et al. (1972)	rat	0.5 Hz, 0.3 - 3 mT, rotating field; exposure of adult animals for 21 - 30 days	Increased ambulatory activity after removal from field
Ossenkopp & Shapiro (1972)	duck egg	0.5 Hz, 2 - 10 and 10 - 30 mT, rotating field; exposure for entire prenatal period	Increased ambulation and defecation rate when tested postnatally

Table 9 (contd).

Reference	Subject	Exposure conditions[a]	Results
deLorge (1972, 1973a,b, 1974, 1979, 1985)	monkey	10, 15, 45, 60, and 75 Hz, 0.8 - 1 mT; fields applied in 4 - 13 daily sessions of 2- to 8-h duration	No consistent influence on motor activity, reaction time, inter-response time, overall level responding, or match-to-sample performance
Beischer et al. (1973)	human being	45 Hz, 0.1 mT; 22.5-h exposure	No effect on reaction time
Gibson & Moroney (1974)	human being	45 Hz, 0.1 mT; 24-h exposure	No consistent effect on cognitive or psychomotor functions
Mantell (1975)	human being	50 Hz, 0.3 mT; 3-h exposure	No effect on reaction time
Medvedev et al. (1976)	human being	50 Hz, 10 - 13 μT; acute exposures	Increased latency of sensorimotor reactions
Smith & Justesen (1977)	mouse	60 Hz, 1.4 - 2 mT; 2-min aperiodic exposures over 2 days	Increased locomotor activity and aggression-related vocalization
Andrianova & Smirnova (1977)	mouse	100 Hz, 10 mT; acute exposures	Heightened motor activity
Brown & Scow (1978)	hamster	10^{-5} Hz, 0.8 - 26 μT; 26-h schedule of high (14 h) to low (12 h) field switching over period of 4 - 5 months	Modified circadian rhythm in locomotor activity
Tucker & Schmitt (1978)	human being	60 Hz, 1.06 mT over whole body, or 2.12 mT over head region; repetitive acute exposures	No perception of field
Becker, von (1979)	termite	50 Hz, 0.05 μT in shielded room; exposures up to several weeks	Stimulation of gallery building activity

Table 9 (contd).

Clarke & Justesen (1979)	chicken	60 Hz, 2.4 mT; aperiodic exposures during 1-h interval for 10 days	Increased variability of response to electric shock stimulus when 60-Hz magnetic field used as conditional stimulus
Udintsev & Moroz (1982)	rat	50 Hz, 20 mT, 15 min/day for 7 days	Transitory stimulation of adrenal system
Udintsev & Moroz (1982)	rat	50 Hz, 20 mT, 6.5 h/day for 7 days	Significant changes in hormone levels
Delgado et al. (1983)	monkey	9 - 500 Hz, 0.1 mT (applied to cerebellum); 9-h daily exposures for maximum of 19 days	Modification of threshold for excitation of motor neurons
Papi et al. (1983)	pigeon	0.034, 0.043, and 0.067 Hz, 60 µT peak intensity; exposures up to 4 h	Initial disturbance of orientation, but no effect on homing performance
Graham et al. (1984)	human being	60 Hz, 40 µT; acute exposures	No perception of field
Creim et al. (1984)	rat	60 Hz; 3.03 mT; 1-h exposure	No field-associated avoidance behaviour
Davis et al. (1984)	mouse	60 Hz; 2.33 mT; 3-day continuous exposure	No change in memory retention, locomotor activity, or sensitivity to a neuro-pharmacological agent
Liboff et al. (1985)	rat	60 Hz, 56 µT (with a transverse 26-µT static field); 30-min exposures	Changes in timing discrimination
Creim et al. (1985)	rat	60 Hz, 3 mT, 1- to 23-h exposures	No avoidance of applied field in a shuttle box test

[a] The magnetic fields were sinusoidal unless otherwise indicated.

ELF electric and magnetic fields (Southern, 1975; Larkin & Sutherland, 1977; Greenberg et al., 1981a,b) have not been included, because of difficulty in attributing these effects to either the electric or magnetic field component. In the case of bees, it appears that ELF electric fields may induce step-potential currents in the hive that have harmful effects when the field intensity exceeds approximately 2 kV/m (Greenberg et al., 1981b). However, altered behavioural patterns in honeybees have also been reported to occur in strong 60-Hz magnetic fields in the absence of an external electric field (Caldwell & Russo, 1968). The mechanism underlying the observed disruption of avian migration by the 72- to 80-Hz electric and magnetic fields from an ELF communication test system is not known (Southern, 1975; Larkin & Sutherland, 1977). However, there are numerous reports that weak static magnetic fields, comparable in strength to the earth's field, may influence the migration patterns of birds (Keeton, 1971; Emlen et al., 1976; Bookman, 1977) and very weak time-varying magnetic fields have also been claimed to affect avian orientation (Papi et al., 1983).

In assessing the effects of time-varying magnetic fields on the behaviour of mammalian species, the publications on this subject, listed in Table 9, are nearly equally divided between positive findings and observations of no behavioural effects in mammals. However, a careful examination of this list leads to the interesting conclusion that most investigations in which behavioural effects were not observed, the time rate of change of the applied magnetic field was sufficient to induce peak intracranial current densities at, or above, the endogenous level of approximately 1 mA/m^2. In contrast, only one of the positive findings of behavioural alterations in mammals (Andrianova & Smirnova, 1977) involved the use of a time-varying magnetic field capable of inducing intracranial currents at this level.

In examining the possible reasons for this apparent disparity, it is important to assess the potential influence on animal behaviour of extraneous factors, such as mechanical vibration and audible noise, that may accompany the activation of magnet coils. The importance of these factors has been well demonstrated by Tucker & Schmitt (1978), who found that perceptive individuals could sense the presence of a 60-Hz magnetic field through auxiliary clues. When these investigators developed an exposure chamber that provided extreme isolation from vibration and audible noise, none of the more than 200 individuals tested could detect 60-Hz fields with intensities of 1.1 mT over the whole body or 2.1 mT over the head region. The sensitivity of behavioural indices to adventitious factors, such as changes in barometric pressure, was also discussed by deLorge (1973b), who emphasized that the correlation of such variables to positive findings of apparent time-varying magnetic field effects must be examined.

6.4 Cellular, Tissue, and Whole Organism Responses

Magnetic fields with a broad range of intensities, ELF frequencies, waveforms, and exposure durations have been evaluated for their ability to induce effects at the cellular, tissue, and animal levels. These studies have recently been reviewed (Tenforde, 1985c, 1986a,d), and only a brief summary will be given here of the cellular and tissue responses to ELF magnetic fields that have been reported on the basis of both in vitro and in vivo studies.

Reports of alterations produced in cellular, tissue, and animal systems as a result of exposure to low frequency magnetic fields are summarized in Table 10, where a brief summary is given of the principal findings in each study. The following types of investigations have not been included in Table 10 for the reasons stated below:

(a) Studies on time-varying magnetic field effects on the visual system (magnetophosphene induction), nervous tissues, and animal behaviour, and epidemiological studies on carcinogenic risk, because these subjects are discussed elsewhere in this section and in section 8;

(b) Reports lacking adequate documentation of field exposure conditions (e.g., frequency, waveform, intensity, and duration of exposure). Similarly, studies in which the biological measurements were qualitative rather than quantitative, as in certain medical reports on bone fracture reunion following therapy with pulsed magnetic fields;

(c) Reports of research that involved combined exposures to ELF electric and magnetic fields, because of the obvious difficulty in delineating the relative effects of the two types of fields.

The reported changes resulting from ELF magnetic field exposure include the following:

(a) Altered cell growth rate (Batkin & Tabrah, 1977; Tabrah et al., 1978; Goodman et al., 1979; Greenebaum et al., 1979, 1982; Aarholt et al., 1981; Ramon et al., 1981; Phillips et al., 1986a);

Table 10. Effects of exposure to time-varying magnetic fields on cells, tissues, and whole animals

Reference	Test specimen	Exposure conditions[a]	Results
Odintsov (1965)	mouse	50 Hz, 20 mT; 6.5-h single exposure or 6.5 h daily for 15 days	Increased resistance to Listeria infection
Druz & Madiyevskii (1966)	rat	3 Hz, 0.1 - 0.8 T, and 50 Hz, 0.05 - 0.2 T; 1-min exposures	Change in hydration capacity of brain, kidney, and liver tissues
Riesen et al. (1971)	guinea-pig brain mitochondria <u>in vitro</u>	60 Hz, 10 mT; 10 - 110-min exposures	No effect on respiration (oxidative phosphorylation)
Riesen et al. (1971)	rat brain synaptosomes <u>in vitro</u>	60 Hz, 5 - 10 mT; 30-min exposure	Decreased uptake of norepinephrine at 0 °C, but not at 10 °C, 25 °C, or 37 °C
Tarakhovsky et al. (1971)	rat	50 Hz, 13 - 14 mT; exposure for 1 month	Changes in serum chemistry, haematocrit, and tissue morphology
Kreuger et al. (1972)	chicken	45 Hz, 0.14 mT, and 60 Hz, 0.12-mT exposure for 1 month	Reduced growth rate in young animals
Ossenkopp et al. (1972)	rat	0.5 Hz, 0.05 - 0.30 or 0.3 - 1.5 mT, rotating field; exposure during entire gestational period	Increased thyroid and testicle weights at 105 - 130 days of age; no change in thymus or adrenal weights compared with controls
Beischer et al. (1973)	human being	45 Hz, 0.1 mT; 22.5-h exposure	Elevated serum-triglycerides; no effects on blood cell counts or other serum chemistry

Table 10 (contd).

DeLorge (1974)	monkey	15 and 45 Hz, 0.82 - 0.93 mT; fields applied in 5 - 8 daily sessions of 2-h duration	No alteration in blood cell counts or serum chemistry (including trigycerides)
Toroptsev et al. (1974)	guinea-pig	50 Hz, 20 mT; 6.5-h single exposure or 6.5 h daily for 24 days	Pathomorphological changes in testes, kidneys, liver, lungs, nervous tissues, eyes, capillaries, and lymphatic system
Udinstev & Moroz (1974)	rat	50 Hz, 20 mT; 1 - 7 days exposure	Increase in adrenal 11-hydroxy corticosteroids
Mizushima et al. (1975)	rat	50 Hz, 0.12 T; 3-h exposure	Anti-inflammatory effects of field on carrageenan-induced oedema and adjuvant-induced arthritis
Beischer & Brehl. (1975)	mouse	45 Hz, 0.1 mT; 24-h exposure	No change in liver-triglycerides
Mantell (1975)	human being	50 Hz, 0.3 mT; 3-h exposure	No haematological changes
Udintsev et al. (1976)	rat	50 Hz, 20 mT; 1-day exposure	Increased lactate dehydrogenase activity and change in distribution in heart and skeletal muscles
Batkin & Tabrah (1977)	mouse neuroblastoma	60 Hz, 1.2 mT; 13-day exposure	Decreased tumour growth rate
Sakharova et al. (1977)	rat	50 Hz, 20 mT; 1-day exposure	Increased catecholamines in tissue
Kartashev et al. (1978)	yeast	0.1 - 100 Hz, 0.025 - 0.40 mT; 20 - 30-min exposure	Changes in rate of anaerobic glycolysis
Kolesova et al. (1978)	rat	50 Hz, 20 mT; single 24-h exposure and 6.5 h daily for 5 days	Development of insulin deficiency

Table 10 (contd).

Reference	Test specimen	Exposure conditions[a]	Results
Tabrah et al. (1978)	*Tetrahymena pyriformis*	60 Hz, 5 - 10 mT; exposures up to 72 h	Cell division delay, reduced growth rate, increased oxygen uptake
Persinger et al. (1978)	rat	0.5 Hz, 0.1 T - 1 mT, rotating field; 10-day exposure	No significant changes in thyroid follicle numbers, mast cells, adrenal and pituitary weights, body weight, or water consumption
Persinger & Coderre (1978)	rat	0.5 Hz, 0.01 T - 1 mT, rotating field; 5-day exposure	No significant change in thymus mast cell numbers in animals exposed prenatally and postnatally or exposed as adults
Udintsev et al. (1978)	rat	50 Hz, 20 mT; 0.25- to 6.5-h and 24 h exposures	Changes in iodine uptake by the thyroid and thyroxine uptake by tissues
Udintsev & Khlynin (1979)	rat	50 Hz, 20 mT; 1-day exposure	Metabolic changes in testicle tissue
Kronenberg & Tenforde (1979)	cultured mouse tumour cells	60 Hz, 2.33 mT; 4-day exposure	No effect on cell growth rate
Chandra & Stefani (1979)	mouse mammary carcinoma	60 Hz, 0.16 T; 1-h daily exposures for 1 - 4 days	No effect on tumour growth rate
Goodman et al. (1979); Greenebaum et al. (1979, 1982)	slime mould	75 Hz, 0.2 mT; 400-day exposure	Lengthened nuclear division cycle and altered respiration rate (decreased O_2 uptake)

Table 10 (contd.)

Chiabrera et al. (1979)	frog erythrocytes	Single bidirectional pulses at 40 - 70 Hz, or 4-kHz bursts of bidirectional pulses with 10 - 20 Hz repetition rate; 2-mT peak intensity; 12- to 24-h exposures	Changes in chromatin structure in the cell nucleus, suggestive of dedifferentiation
Kolodub & Chernysheva (1980)	rat	50 Hz, 9.4, and 40 mT; 5 h daily for 15 days	Altered brain metabolism at higher field intensity, including decreased rate of respiration, decreased levels of glycogen, creatine phosphate and glutamine, and increased DNA content
Fam (1981)	mouse	60 Hz, 0.11 T; 23 h daily for 7 days	Decreased body weight and increased water consumption; haematology, organ histology, and reproduction not affected
Aarholt et al. (1981)	bacteria	16.66 and 50 Hz, 0 - 2 mT; 10- to 12-h exposure	Decreased growth rate
Ramon et al. (1981)	bacteria	60 and 600 Hz, 2 mT; 17- to 64-h exposure	Decreased growth rate and cytolysis
Toroptsev & Soldatova (1981)	rat	50 Hz, 20 mT; 1- to 24-h exposure	Pathomorphological changes in brain
Kolodub et al. (1981)	rat	50 Hz, 9.4 - 40 mT, daily 3-h exposures for up to 6 months	Changes in carbohydrate metabolism in the myocardium
Sakharova et al. (1981)	rat	50 Hz, 20 mT, 1-day exposure	Changes in catecholamine content and morphology in brain, heart, liver, spleen, and circulatory system

Table 10 (contd.)

Reference	Test specimen	Exposure conditions[a]	Results
Delgado et al. (1981, 1982)	chicken embryo	10, 100, and 1000 Hz; 0.12, 1.2, and 12 μT; 0.5-ms rectangular pulses; 2-day exposure	Morphological abnormalities in nervous tissue, heart, blood vessels, and somites
Soldatova (1982)	rat	50 Hz; 20, 40, and 70 mT; 6.5 h daily for 5 days, or 24-h continuous exposure	Pathomorphological changes in brain tissue
Sander et al. (1982)	human being	50 Hz, 5 mT; 4-h exposure	No changes in ECG, EEG, hormones, blood cell counts, or blood chemistry
Luben et al. (1982)	mouse osteoblast culture	Single bidirectional pulses at 72 Hz, or 4-kHz bursts of bidirectional pulses with 15 Hz repetition rate; 2 mT peak intensity; 3-day exposure	Reduced cAMP production in response to parathyroid hormone
Shober et al. (1982)	mouse	10 Hz, 1 mT; 1-day exposure	Decreased sodium ion content of liver
Norton (1982)	cultured chicken embryo sternum	4 kHz bursts of bidirectional pulses with 15 Hz repetition rate; 2 mT peak intensity; four 6-h exposures during 2 days	Increased hydroxyproline, hyaluronate, and DNA synthesis; decreased glycosoaminoglycans; increased lysozyme activity
Aarholt et al. (1982)	bacteria (*E. coli*)	Square wave pulses at 50 Hz; 0.20 - 0.66 mT, 2- to 3-h exposures	Changes in rate of β-galactosidase synthesis
Dixey & Rein (1982)	rat pheochromocytoma cell *in vitro*	500 Hz; bidirectional pulses; 160 - 850 μT; 3-h exposure	Stimulation of noradrenaline release

Table 10 (contd.)

Conti et al. (1983)	cultured human lymphocyte	1, 3, 50, and 200 Hz; 2.3 - 6.5 mT; square-wave pulses; 3-day exposure	Inhibition of lectin-induced mitogenesis by 3- and 50-Hz fields
Goodman et al. (1983)	sciara-coprophila salivary giant chromosome	Single bidirectional pulses at 72 Hz, or 4-kHz bursts of bidirectional pulses with 15 Hz repetition rate; 2 mT peak intensity; 5- to 90-min exposures	Initiation and increase of RNA transcription at defined loci
Jolley et al. (1983)	rabbit pancreas	4 kHz bursts of bidirectional pulses with 15-Hz repetition rate; 2-mT peak intensity; 18-h exposure	Reduced Ca^{++} content and efflux; reduced insulin release during glucose stimulation
Ramirez et al. (1983)	Drosophila egg	0.5-ms square-wave pulses at 100 Hz, 1.76 mT peak-to-peak intensity; or 50 Hz, 1.41 mT sinusoidal field; 2-day exposure	Decreased viability of eggs
Ubeda et al. (1983)	chicken embryo	0.5 ms bidirectional pulses at 100 Hz (4 different waveforms); 0.4- to 104-μT peak intensity; 2-day exposure	Teratogenic changes in nervous system, circulatory system, and foregut
Archer & Ratcliffe (1983)	cultured chicken tibia	1 Hz, 15 - 60 mT square-wave pulses; 7-day exposure	Decreased collagenous and non-collagenous protein synthesis; no alteration in glycosoaminoglycan DNA synthesis
Liboff et al. (1984)	cultured human fibroblast	15 Hz - 4 kHz; 2.3 - 560 μT; 18- to 96-h exposures	Increased DNA synthesis
Cain et al. (1984)	cultured mouse calvarium	Single bidirectional pulses at 72 Hz, or 4-kHz bursts of bidirectional pulses with 15-Hz repetition rate; 2.5-mT peak intensity; exposure for 1 - 16 h	Inhibition of cAMP production and Ca^{++} release in response to parathyroid hormone

Table 10 (contd).

Reference	Test specimen	Exposure conditions[a]	Results
Temur'yants et al. (1985)	rat	8 Hz, 5.2 μT; daily 3-h exposures for up to 45 days	Transient hyperlipaemia in blood-serum
Murray & Farndale (1985)	cultured chicken fibroblast	15-Hz bidirectional pulses; 2.2-mT peak intensity; daily 12-h exposures for 1 - 8 days	Enhanced collagen and total protein synthesis, and decreased cAMP after 6 days of exposure
Cain et al. (1985)	cultured mouse calvarium	Single bidirectional pulses at 15 Hz; 0.8-mT peak intensity; 15- to 60-min exposures	Decreased cAMP production and increased ornithine decarboxylase activity in response to parathyroid hormone
Ueno et al. (1985)	toad embryo (*Xenopus laevis*)	20 Hz, 2 kHz, and 20 kHz; 10 - 15 mT; 15-min to 8-h exposures	Teratogenic effects
Gundersen & Greenebaum (1985)	rat muscle	60 and 70 Hz (linear and circular polarization); 0.1 mT; 10-min exposure	Effects on minature endplate potentials
Winters et al. (1985a,b)	human and dog leukocytes	60 Hz, 0.1 mT; 24-h exposure	No effects on mitogen responses, DNA, RNA or protein synthesis, or levels of cell surface receptors
Phillips et al. (1986a,b)	cultured human colon tumour	60 Hz, 0.14 mT; 1-day exposure	Increase in growth rate, number transferring receptors, and expression of tumour-specific antigens

[a] The magnetic fields were sinusoidal unless otherwise indicated.

(b) Decreased rate of cellular respiration (Cook et al. 1969; Goodman et al., 1979; Greenebaum et al., 1979, 1982; Kolodub & Chernysheva, 1980);

(c) Altered metabolism of carbohydrates, proteins, and nucleic acids (Udintsev et al., 1976, 1978; Kartashev et al., 1978; Udintsev & Khlynin, 1979; Kolodub & Chernysheva, 1980; Kolodub et al., 1981; Norton, 1982; Archer & Ratcliffe, 1983; Buyavikh, 1984; Liboff et al., 1984);

(d) Effects on gene expression and genetic regulation of cell function (Chiabrera et al., 1978, 1979; Beltrame et al., 1980; Aarholt et al., 1982; Goodman et al., 1983; Goodman & Henderson, 1986);

(e) Endocrine alterations (Riesen et al., 1971; Udintsev & Moroz, 1974; Sakharova et al., 1977, 1981; Kolesova et al., 1978; Udintsev et al., 1978);

(f) Altered hormonal responses of cells and tissues, including effects on cell surface receptors (Dixey & Rein, 1982; Luben et al., 1982; Marsakova, 1983; Jolley et al., 1983; Cain et al., 1984; Chan & Nicholson, 1986);

(g) Altered immune response to antigens and mitogens (Odintsov, 1965; Mizushima et al., 1975; Conti et al., 1983; Budd & Czerski, 1985);

(h) Morphological and other nonspecific tissue changes in adult animals, frequently reversible with time after exposure (Druz & Madiyevskii, 1966; Toroptsev et al., 1974; Sakharova et al., 1981; Toroptsev & Soldatova, 1981; Soldatova, 1982; Shober et al., 1982);

(i) Teratological and developmental effects (Ossenkopp et al., 1972; Delgado et al., 1981, 1982; Kreuger et al., 1972; Ramirez et al., 1983; Ubeda et al., 1983).

These observations were made with sinusoidal and square-wave time-varying magnetic fields and with pulsed magnetic fields that had repetition rates in the ELF frequency range. With few exceptions, the peak field intensities that were used exceeded 0.5 mT and the current density induced in the exposed samples exceeded 10 mA/m². The currents induced within the cellular and tissue fluids were therefore at, or above, the upper limit of the naturally occurring levels.

It is noteworthy that most of the studies with square waveforms and with pulsed fields that induced current densities greater than 10 mA/m^2 led to findings of positive bioeffects (Delgado et al., 1981; Dixey & Rein, 1982; Luben et al, 1982; Norton, 1982, Archer & Ratcliffe, 1983; Conti et al., 1983; Goodman et al., 1983; Jolley et al., 1983; Ramirez et al., 1983; Cain et al., 1984). Developmental effects were observed at lower induced current density levels by Delgado et al. (1982) and Ubeda et al. (1983), when they exposed chick embryos to pulsed magnetic fields. Juutilainen et al. (1986) and Juutilainen & Saali (1986) found that this effect depended on the waveform and frequency of the magnetic field. A large international study is now in progress in an effort to replicate these findings. It has been suggested that the currents induced by such fields could exert an electrochemical effect at the cell surface (Luben et al., 1982; Jolley et al., 1983). This effect, in turn, may influence hormone-receptor interactions, adenylate cyclase activity, and the membrane transport and intracellular concentration of calcium ions. All of these membrane functions are known to play an important role in cell metabolism and growth dynamics.

Aarholt et al. (1982) measured the rate of beta-galactosidase synthesis in cultures of E. coli exposed to 50-Hz square-wave magnetic fields, in order to investigate the effect of such exposure on the lac operon function. Following a 30-min exposure at 0.2 mT - 0.30 mT, a decrease in beta-galactosidase synthesis rate of about one-third was reported. At 0.32 mT, the synthesis rate returned to control values, and increased by a factor of 2 at 0.54 and 0.56 mT. No differences compared with control values was seen at 0.58 mT and higher values up to 0.70 mT. No measurements were made at higher field strengths.

Chiabrera et al. (1979) reported a decrease in the chromatin density of frog erythrocytes exposed to pulsed magnetic fields, such as those used in bone growth stimulation. This imparted to the cells an appearance of earlier maturation stages. There were morphological and cytophotometric changes in chromatin density, which suggested gene depression, but such a conclusion does not appear to be justified, since RNA, protein, and/or haemoglobin synthesis were not investigated.

Using biochemical and autoradiographic techniques, Goodman et al. (1983) demonstrated the initiation of RNA transcription at two different sets of loci in salivary gland giant chromosomes exposed to pulsed magnetic fields. One set of loci became activated following 45 min of exposure to single pulses with a 72-Hz repetition rate, another set after 15 min of exposure to pulse trains with a repetition rate of 15 Hz. Changes in protein synthesis in salivary gland cells exposed under identical conditions, reported by Ryaby et al. (1983), offer confirmatory evidence. All the reports quoted above seem to indicate that pulsed magnetic fields may affect gene expression. However, it should be noted that these studies were not duplicated or otherwise verified by independent teams of research workers.

Eighteen of the investigations with ELF sinusoidal magnetic fields have involved exposure of rodents to 50-Hz and 60-Hz fields with intensities ranging from 0.01 to 0.8 T (Odintsov, 1965; Druz & Madiyevskii, 1966; Tarakhovsky et al., 1971; Toroptsev et al., 1974; Udintsev & Moroz, 1974; Mizushima et al., 1975; Udintsev et al., 1976; Sakharova et al., 1977, 1981; Kolesova et al., 1978; Udintsev et al., 1978; Udintsev & Khlynin, 1979; Chandra & Stefani, 1979; Kolodub & Chernysheva, 1980; Fam, 1981; Kolodub et al., 1981; Toroptsev & Soldatova, 1981; Soldatova, 1982). With the exception of one report in which tumour growth rate was observed not to be influenced by brief exposure to a 60-Hz, 0.16-T field (Chandra & Stefani, 1979), all of the studies report positive findings of cellular and tissue effects from ELF magnetic fields. The maximum current densities induced in the experimental animals by the applied field exceeded approximately 10 mA/m^2 in these studies, and were therefore at, or above, the upper limit of the endogenous currents that are normally present within the body (Bernhardt, 1979).

In contrast to the findings of positive biological effects listed above, present evidence suggests that animal haematological parameters are unaffected by ELF magnetic fields at intensities that reportedly influence other cellular and tissue systems. With the exception of one isolated report (Tarakhovsky et al., 1971), all of the published studies on haematological parameters in exposed animals have shown no consistent field-associated effects (Beischer et al., 1973; deLorge, 1974; Mantell, 1975; Goldberg & Mel'nik-Guykazyan, 1980; Fam, 1981; Sander et al., 1982). The apparent lack of sensitivity of the haematological system to magnetic fields is in distinct contrast to the well-documented effects of ionizing radiation and high-intensity microwave fields on this particular physiological system.

Three of the studies listed in Table 10 involved short-term exposures of human volunteers to ELF magnetic fields (Beischer et al., 1973; Mantell, 1975, Sander et al., 1982). With the exception of one unconfirmed report of an elevation in serum-triglycerides in the exposed subjects (Beischer et al., 1973), none of these investigations revealed adverse effects of ELF magnetic fields with intensities comparable to or exceeding the levels generally encountered by man. Particularly notable in this regard is the report by Sander et al. (1982), who observed that a 4-h exposure of human volunteers to a 50-Hz, 5-mT field produced no changes in serum chemistry, blood cell counts, blood gases and lactate concentration, electrocardiogram, pulse rate, skin temperature, hormones (cortisol, insulin, gastrin, thyroxine), and various neuronal measurements, including visually evoked potentials recorded in the electroencephalogram.

6.5 Effects of Pulsed Magnetic Fields on Bone Growth and Repair

Direct current electrical stimulation has been used since the nineteenth century for the treatment of bone non-unions and pseudarthroses. Although this procedure has met with some success clinically, the use of direct currents has been shown to produce several undesirable side-effects including:

(a) surgical trauma and a risk of infection through the implantation of electrodes in bone;

(b) the development of electrode polarization with time, which leads to increased impedance and decreased current for a given applied voltage;

(c) osteogenesis, which has been found to increase near the negative electrode (cathode), but decrease near the positive electrode (anode).

These disadvantages of direct current electrical stimulation have been overcome by the recent introduction of pulsed magnetic field generators as a means of inducing ELF electrical currents within bone tissue (Bassett et al., 1974). By using magnetic coils placed about a limb containing a fractured bone, electric fields with a typical strength of 0.2 - 2 V/m can be induced within the bone tissue. In the usual configuration, two coils are placed about the limb and positioned such that the bone fracture lies along a line joining the centres of the coils, and hence along the magnetic field lines. Assuming the conductivity of bone to be 0.01 S/m at ELF frequencies (Lunt, 1982), the local current densities induced in bone by the pulsed magnetic fields can be estimated to lie in the range of approximately 2 - 20 mA/m². Initial studies on bone fracture reunion in dogs demonstrated that a pulse repetition frequency of 65 Hz was more effective than 1 Hz (Bassett et al., 1974), and several subsequent studies have revealed that frequencies of 60 - 75 Hz are the most advantageous in facilitating fracture union and preventing pseudo-arthroses (Bassett, 1982).

Following the initial demonstration of the efficacy of pulsed magnetic fields in achieving bone fracture reunion in experimental animals, several successful clinical trials have been reported concerning the treatment of bone fractures and arthroses in human beings by this method. In a four-year clinical trial involving more than 100 patients, Bassett et al. (1977) reported an 85% success rate in the treatment of long-established pseudo-arthroses. The successful use of pulsed magnetic fields in the facilitation of bone healing in human subjects has subsequently been reported by several clinical groups (Watson & Downes, 1978; Bassett et al., 1982; Hinsenkamp, 1982; Bigliani et al., 1983).

Barker et al. (1984) recently published an interim report on a double-blind clinical trial in which 9 patients with non-united tibial fractures were treated with active magnetic stimulators, while a group of 7 control patients were fitted with dummy stimulators. After 24 weeks of treatment, the fracture united in 5 of the 9 patients with active stimulators, and fractures in 5 of the 7 patients with dummy stimulators also united. Thus, there was no statistically significant difference between the treated and control groups. This preliminary result suggests that earlier claims of clinical success with pulsed magnetic field applicators may have been biased by the use of control groups that were not subjected to the same immobilization procedure as the patients undergoing active treatment. Controlled, double-blind studies on large numbers of patients are needed to assess this modality of treatment.

The mechanism by which the weak ELF electric currents induced in bone tissues by pulsed magnetic fields could exert an influence on fracture repair is also under investigation in a number of laboratories. Evidence from in vitro studies on osteoblasts and chondrocytes indicates that the pulsed fields influence hormone binding to receptors at the cell surface, and thereby depress the intracellular concentration of calcium ions and cyclic AMP (Bassett, 1982; Luben et al., 1982). These effects, in turn, can significantly influence cellular metabolism and stimulate growth. Studies by Hinsenkamp & Rooze (1982) with in vitro cultures of limbs from mouse fetuses demonstrated that electromagnetic stimulation leads to chondrocyte proliferation and an improved alignment of trabeculae and cartilage. Archer & Ratcliffe (1983) reported that cultured tibias from chicken embryos had a reduced collagen content following exposure to a pulsed magnetic field for 7 days. The observation was also made by these workers that the total synthesis of sulfated glycosoaminoglycans, which are major components of the extracellular matrix, was not affected by exposure to the pulsed magnetic field. The further elucidation of the macromolecular and developmental changes that accompany the stimulation of bone tissue by pulsed ELF magnetic fields remains a challenging area of research, which will ultimately lend useful insight into the mechanisms by which weak ELF fields interact with living cells.

6.6 Conclusions

A well established and repeatable effect of human exposure to ELF magnetic fields is the induction of magnetophosphenes. This effect shows a strong frequency dependence on flux density. The threshold for magnetophosphenes is between 2 and 10 mT in the frequency range of 10 - 100 Hz.

Much more intense fields are required to directly stimulate nerve and muscle tissue. These effects are also frequency dependent with thresholds above 50 mT (10 - 100 Hz).

Numerous investigations with ELF magnetic fields with sinusoidal, square-wave, and pulsed waveforms have led to reports of alterations in cell, tissue, and animal systems, when the induced current density exceeded approximately 10 mA/m^2. These reported changes have included alterations in cell metabolism and growth properties, gene expression, endocrine and immune functions, and teratological and developmental effects. However, several of these studies have not been successfully replicated.

A large number of laboratory studies have revealed evidence of changes in cellular metabolism and growth properties as a result of exposure to pulsed magnetic fields. However, in clinical applications of these fields for the facilitation of bone fracture reunion, not enough double-blind studies on large numbers of patients have been carried out to assess the efficacy of this treatment.

7. HUMAN STUDIES

Since epidemiological studies have assumed an important role in the assessment of the human health risks of non-ionizing radiation exposure, the characteristics of these studies must be considered relevant to determining causal relationships. Although there are inherent limitations in an observational method, sufficient data can be compiled from epidemiological studies to establish a causal relationship, as has occurred, for example, for cigarette smoking and lung cancer.

The term causality is used when there is a biological association, and where a statistical pattern can be inferred. In general terms, a causal relationship is supported by a strong association between exposure and disease. Consistency in demonstrating the same association across different populations, for example different occupational groups or different regions of the country, supports a causal relationship. Exposure to the physical factor prior to the effect is absolutely necessary for the association to be interpreted as causal. A dose-response relationship in which risk shows a positive correlation with a level of exposure provides a stronger inference of causality. Although the mechanism involved does not need to be known exactly, it is highly desirable to develop a predictive theory.

7.1 Studies on Working Populations

7.1.1 Workers exposed to static magnetic fields

Studies on Soviet workers involved in the manufacture of permanent magnets indicated various subjective and physiological symptoms: irritability, fatigue, headache, loss of appetite, bradycardia, tachycardia, decreased blood pressure, altered EEG, itching, burning, and numbness (Vyalov et al., 1964; Vyalov & Lisichkina, 1966; Vyalov, 1967). The strength of the magnetic fields causing these symptoms was not reported and there was no control group, which significantly reduces the value of the reports. A later study on workers in industries involving magnet production and machine building (Vyalov, 1971, 1974), involving 645 exposed persons and 138 controls, reported subjective complaints and minor physiological effects, especially in haematological and cardiovascular indices. The average static magnetic field strengths to which these workers were exposed were typically 2 - 5 mT at the level of the hands and 0.3 - 0.5 mT at the chest and head levels. Unfortunately, no statistical analyses were performed.

Marsh et al. (1982) studied workers (320 exposed, 186 controls) in the USA employed in industries using electrolytic cells that generated large static magnetic fields. The exposed workers were subjected to average magnetic fields of 7.6 mT in operator accessible locations and maximum fields of 14.6 mT. The time-weighted average field exposures were calculated to be 4 and 11.8 mT for the mean and maximum field levels, respectively. Although no major health effects were found, minor haematological alterations and blood pressure changes were observed.

The prevalence of 19 common diseases was studied in 792 workers in high-energy accelerator laboratories, bubble chambers, calutrons (isotope separation facilities), and high-field magnet facilities, compared with the same number of matched controls (Budinger et al., 1984b). A subgroup of 198 workers exposed to 0.3 T or higher static fields for 1 h or longer was also compared with matched controls. No significant changes were found in the prevalence of diseases of the skin; circulation; respiratory tract; male genital organs; genito-urinary tract; bone, muscle, and tendon; gastro-intestinal tract; nervous system; liver and gall bladder; blood; and eye. The prevalences of benign and malignant diseases, allergic and metabolic diseases; senility and other ill-defined diseases; and accidents including poisonings were also unaffected.

In a study on 211 contact welders in the USSR, Abramovich-Poljakov et al. (1979) showed an increase in nervous system disorders and leukocyte counts, and alterations in ECG, compared with 113 non-welders. Although the authors related this to exposure to 0.1- to 0.2-s pulsed magnetic fields of strengths 1000 - 100 000 A/m (1.25 mT - 125 mT), exposure to other hazards, such as metal fumes could also be expected to lead to effects on health.

Milham (1979, 1982, 1985b) reported that workers in the aluminium industry have a significantly elevated mortality from all classes of leukaemia and from acute leukaemia. This conclusion was based on a study of the death records of 438 000 males in the state of Washington (USA) from 1956-79. The proportionate mortality ratios (PMRs) for all classes of leukaemia and acute leukaemia among aluminium workers were 189 and 258, respectively ($\underline{P} < 0.01$). This finding was subsequently confirmed by Rockette & Arena (1983), though their broader study involving 14 aluminium plants in the USA showed only a small overall excess of leukaemia mortality with a standardized mortality ratio of 127.9, which was not statistically significant. The study by Rockette & Arena (1983) also revealed a trend towards increased pancreatic cancer, lymphohaematopoietic cancers, genito-urinary cancer, non-malignant respiratory disease, and various unspecified benign neoplasms. Overall, the elevated risk of these various cancers was not statistically significant. Milham (1982) suggested that the elevated risk of leukaemia among aluminium workers might be associated with exposure to the static magnetic fields that result from the high DC electric currents used in the electrolytic reduction of alumina to aluminium metal. However, at present, there is no clear evidence indicating a link between the magnetic fields present in aluminium plants and the increased incidence of leukaemia or other cancers. The process used for aluminium production creates coal-tar pitch volatiles, fluoride fumes, sulfur oxides, and carbon dioxide. All of these environmental contaminants must be taken into account in any attempt to relate magnetic field exposure and cancer risk among workers in the aluminium industry.

Two other recent studies on persons exposed occupationally to static magnetic fields have failed to detect an elevated risk of cancer (Budinger et al., 1984b; Barregard et al., 1985). The results of the study by Budinger et al. (1984b) did not reveal any elevation in the incidence of benign or malignant neoplasms among 792 exposed workers compared with an equal number of matched controls. Barregard et al. (1985) studied cancer incidence during a 25-year period among a small cohort of workers at a chloroalkali plant where the 100-kA DC currents used for the electrolytic production of chlorine gave rise to static magnetic fields of 4 - 29 mT in the working environment. The observed versus expected incidence of cancer among these workers was not significantly different.

Some of the reported effects in man exposed to magnetic fields are summarized in Table 11. Although these studies are inconclusive, they suggest that, if long-term effects occur, they are very subtle, since no cumulative gross effects are evident. In general, the available data on cancer incidence among workers in occupations that involve exposure to large static magnetic fields do not support an association between cancer incidence and exposure to these fields.

7.1.2 Cancer epidemiological studies on workers exposed to ELF electromagnetic fields

Preliminary observations, some published as letters to the editor (Milham 1982; Wright et al., 1982; McDowall, 1983; Vagerö & Olin, 1983; Coleman et al., 1983; Gilman et al., 1985; Lin et al., 1985; Milham, 1985a,b; Pearce et al., 1985; Stern et al., 1986) reported an epidemiological association of leukaemia and other tumours with electrical/electronic occupations involving presumed exposure to power-frequency electromagnetic fields (Table 12).

Table 11. Studies of workers exposed to static magnetic fields

Exposure characteristics	Reported effects (exposed population)	Reference
Workers in magnet production; average exposure: 2 - 5 mT (hands), 0.3 - 0.5 mT (chest and head)	Subjective and minor physiological effects (645 exposed, 138 controls, no statistical analysis)	Vyalov (1974)
Contact welders; 0.1- to 0.2-s pulsed magnetic fields of 1.25 - 125 mT, 8 h/day	Increased nervous system, cardiac, and blood disorders (211 exposed, 113 controls)	Abramovich-Poliakov et al. (1979)
Workers in aluminium plants (no fields reported)	Increased risk of leukaemia (death records of 438 000 males, but few cases)	Milham (1979, 1982, 1985b)
Industries using electrolytic cells (average, 7.6 mT; maximum, 14.6 mT)	Minor haematological alterations, but no major health effects (320 exposed, 186 controls)	Marsh (1982)
Workers in aluminium plants (no fields reported)	Small excess of leukaemia mortality; non-significant risk of other cancers	Rockette & Arena (1983)
High energy accelerator laboratory (fields up to 2 T)	No increased prevalence of 19 common diseases including cancers (792 exposed, 792 controls)	Budinger et al. (1984b)
Electrolytic production of chlorine (fields 4 - 29 mT)	No increased incidence of cancer over 25-year period	Barregard et al. (1985)

In an analysis of data for occupational mortality, Milham (1982) noted higher than expected proportionate mortality due to acute myeloid leukaemia among men "whose occupation requires them to be in electric or magnetic fields." The data base consisted of 438 000 deaths of men, 20 years of age or older who, from 1950 to 1979, were residents of Washington state (USA). Although the proportionate mortality ratio (PMR = observed/expected x 100) is a useful statistical measure, it has technical limitations that should be explored in a complete study. PMRs significant at the $\underline{P} < 0.01$ level were observed for "electricians", TV and radio repairmen, power-station operators, and aluminium workers, though similarity in field exposure among these groups was not proved and is unlikely.

Table 12. Cancer incidence and occupational exposure to power frequency electromagnetic fields

Reference	Subject	Cancer risk
Wiklund et al. (1981)	Telecommunication workers	No cancer risk
Milham (1982, 1985b)	Electrical occupations	Increased leukaemia
Wright et al. (1982)	Electrical occupations	Increased leukaemia
McDowall (1983)	Electrical occupations	Increased leukaemia
Coleman et al. (1983)	Electrical occupations	Increased leukaemia
Vagerö & Olin (1983)	Electrical occupations	No leukaemia risk
Swerdlow (1983)	Electrical occupations	Increased eye melanoma
Pearce et al. (1985)	Electrical occupations	Increased leukaemia
Lin et al. (1985)	Electrical occupations	Increased brain tumours
Milham (1985a)	Amateur radio operators	Increased leukaemia
Gilman et al. (1985)	Males in underground mines	Increased leukaemia
Vagerö et al. (1985)	Electrical occupations	No leukaemia risk; increased urinary cancer; increased malignant melanoma
Calle & Savitz (1985)	Electrical occupations	No leukaemia risk
Olin et al. (1985)	Electrical occupations	Increased malignant melanoma
Stern et al. (1986)	Electrician and welders	Increased leukaemia
Tornqvist et al. (1986)	Electric power industry	No leukaemia risk; no brain tumour risk

Wright et al. (1982) sought to verify Milham's (1982) results by examining a similar statistic, the proportional incidence ratio (PIR) of a different and much smaller data base. They found significant increases ($\underline{P} < 0.05$) in the incidence of acute myeloid leukaemia, based on a total of 4 cases in power linemen and telephone linemen, two groups for which the Washington data yield insignificant PMRs. Calle & Savitz (1985) analysed mortality from leukaemia among 81 men in electrical occupations in Wisconsin during the period 1963-78. The classification of occupational groups used by these authors was identical to those of Milham (1982) and Wright et al. (1982). PMR was calculated on the basis of all deaths occurring during this period in Wisconsin. No excess mortality from leukaemia was found, with the possible exception of acute leukaemia in electrical engineers. The PMR was 257 (one-sided $P < 0.05$). When the leukaemia mortality data were pooled across all 10 electrical occupations, the PMR values were 103 and 113 for all leukaemia and acute leukaemia, respectively. Calle & Savitz (1985) concluded, on this basis, that there was no consistent overall pattern of excess leukaemia risk among workers in electrical occupations.

Additional data on occupational leukaemia rates in the United Kingdom were provided in two letters to the editor. McDowall (1983) found increased evidence of leukaemia in occupationally exposed electrical workers using PMRs and also by a case-control study. Coleman et al. (1983) also examined the leukaemia incidence for the same electrical occupations with evidence for a 17% excess that was especially strong for electrical fitters and telegraph operators, for whom the extent of electric or magnetic field exposure has not been established.

The suggestion of a small, but significant, increase in the risk of leukaemia in electrical workers in New Zealand with the potential for exposure to alternating electrical and magnetic fields was found by Pearce et al. (1985). The authors stated that their study would also support that the increased risk of leukaemia was due to exposure to metal fumes and substances used in electrical component assembly, since the greatest excess of risks was found for electronic equipment assemblers and radio and television repairers.

A recent study by Stern et al. (1986) has led to the observation of an elevated incidence of leukaemia among electricians and welders in the Portsmouth Naval Shipyard (New Hampshire, USA). A matched case-control study was conducted of 53 leukaemia deaths and 212 controls identified from a population of 24 545 workers employed at this naval nuclear shipyard between 1 January 1952 and 15 August 1977. No correlations were found between leukaemia mortality and exposure to ionizing radiation or to organic solvents. The Mantel-Haenszel odds ratio was 3 for the mortality from lymphatic leukaemia among the electricians ($P < 0.05$). For welders, the odds ratio was 2.25 for myeloid leukaemia ($P < 0.05$). These elevations in leukaemia mortality were attributed by the authors to electromagnetic field exposure among workers in the affected groups.

Other studies on groups with presumed occupational exposure to electromagnetic fields have failed to detect an excess of leukaemia cases (Vagerö & Olin, 1983; Vagerö et al., 1985; Tornqvist et al., 1986). However, in these studies, a significant increase in the incidence of pharyngeal cancer (Vagerö & Olin, 1983), urinary cancer (Vagerö et al., 1985), and malignant skin melanoma (Olin et al., 1985; Vagerö et al., 1985) was noted. An excess risk of malignant melanoma of the skin was primarily associated with occupations that involved soldering.

Using years of employment as a measure of exposure to electromagnetic fields, Gilman et al. (1985) reported a significant increase in the incidence of leukaemia among white male coal miners who had worked for more than 25 years underground compared with miners who had worked for less than 25 years underground. It was suggested that the electromagnetic fields associated with power lines, transformers, etc. were a possible factor in this increased risk.

In an epidemiological study on telecommunications workers based on the Swedish Cancer Environment Registry, Wiklund et al. (1981) did not find any increased risk for this occupational group compared with the Swedish population as a whole.

Swerdlow (1983) suggested an association with an increase in the incidence of adult melanoma of the eye in electrical and electronic workers and also the non-manual social classes (white-collar workers).

An increased incidence of cancer deaths in male members of the American Radio Relay League in California and Washington States was found by Milham (1985a). Lin et al. (1985) recently reported an increased number of brain tumour deaths among white male workers in 3 electrical/electronic occupations in the state of Maryland (USA) during the period 1969-82. The Mantel-Haenszel odds ratio was 2.15 (with a 95% confidence interval of 1.10 - 4.06) for workers who had experienced definite electromagnetic field exposure during the course of their work.

7.1.3 Conclusions

The association between cancer incidence and occupational exposure to power-frequency electric and magnetic fields suggested by many of the recent epidemiological studies reviewed here is not clearly consistent. In many of these studies, the ELF field levels to which the occupational groups under study were exposed were not characterized. Also, in a number of the investigations, confounding variables of high carcinogenic potential, e.g., certain organic fumes and hydrocarbon particulates, were not taken into account. Therefore, even if it is concluded that the risk of leukaemia or other types of cancer was increased for certain occupational groups, it does not follow that the ELF electric or magnetic field exposure was the relevant etiological factor.

In general, given the limited statistical power of the studies reported to date, the reported increase in the incidence of leukaemia and other cancers has been less than a factor of 2 (for example, from 1 per 10^6 to 2 per 10^6) compared with a case-control group or the general population. These epidemiological studies have often involved very few disease cases in an occupational category, as well as inconsistent category definitions. As discussed in the introduction to this section, epidemiological methods can detect associations with a reasonable degree of certainty in studies such as these, if appropriate criteria are applied to a large enough data base of good integrity. The suggestion of leukaemia and other cancers related to ELF electromagnetic field exposure raises important questions that should be addressed by studies of adequate statistical power, in which confounding variables are taken into account. There is an urgent need for well-designed experimental studies on the carcinogenic effects of ELF electromagnetic field exposure, using the time-honoured methods that have been previously used for testing the carcinogenic effects of chemical substances. Until such data are obtained and additional epidemiological studies are carried out, the problem of the carcinogenic effects of ELF electromagnetic field exposure should be considered to be unresolved.

7.2 Epidemiological Studies on the General Population

Wertheimer & Leeper (1979) reported a 2- to 3-fold increase in the incidence of leukaemia among Colorado children, presumably exposed to fields from high electric current configurations. Magnetic fields (associated with the electric currents) were estimated by scoring the type of electrical wiring configuration close to the homes into categories of high or low current configurations.

The same authors (Wertheimer & Leeper, 1982) extended their work to a study of the incidence of adult cancer in those living near high-current electric wiring. The associations demonstrated were not dependent on age, urbanicity, neighbourhood or socio-economic level and were most clearly demonstrated where urban/industrial factors were not present to obscure the pattern. The four types of cancer that appeared to be particularly elevated in the exposed adult populations were cancer of the nervous system, uterus, breast, and lymphomas. The authors suggested that magnetic fields might have a tumour-promoter effect, since the increases were maximal at 7 years from the time of taking up residence in the area.

These preliminary studies have limitations common to many epidemiological studies involving cohort selection and additional problems suggesting possible biases in the techniques for scoring the wiring configurations, and in the assumption that the scoring technique accurately determines magnetic field strength levels among the cases examined. Further questions are raised, because cases were ascertained after death, and therefore no account was taken of cancer cases still alive and, because birth and death addresses were used, again introducing the potential for observer bias. Considerable interest has been provoked by these findings and it is expected that many of the issues will be dealt with in future research.

The hypothesis that such weak magnetic fields (of the order of 0.1 - 0.7 μT) produce biological effects has raised questions, such as those of Miller (1980), who criticized the Wertheimer & Leeper (1979) study on the basis that the magnetic field from electrical appliances in the home would far exceed contributions from electrical wiring configurations in the environment.

Tomenius et al. (1982) and Tomenius (1986) reported an increased incidence of tumours (malignant and benign) in children living in homes where the magnetic field outside the front door was more than 0.3 μT. The data involved a small number of cases and again the field measurement was questionable, because the relation of personal exposure to the value of the field measured outside the home was not established. Tomenius (1986) did not find an increased incidence of leukaemia but an increased incidence of nervous system tumours in residences with magnetic fields greater than 0.3 μT. Furthermore, if a cut-off magnetic field strength other than 0.3 μT was used, no association of tumour incidence and magnetic field exposure would occur.

These studies, and the preliminary occupational data (see above) causing some concern in relation to electric or magnetic field exposure, must be investigated further to determine whether the suggested link with cancer induction or promotion can be established. Recently, the results of three studies carried out in the United Kingdom did not show any association between magnetic fields and cancer (Coleman et al., 1985; Myers et al., 1985; McDowall 1986). It should be noted that these studies are open to the same criticisms as those above that indicate an association, particularly with regard to the limited statistical power and lack of quantification of exposure. A summary of studies on cancer incidence and population exposure to electromagnetic fields is given in Table 13.

Another aspect of ELF magnetic field effects that should be considered in the context of behavioural alterations is the report of a correlation between the incidence of suicides and the intensity of residential 50-Hz magnetic fields from power-line sources (Perry et al., 1981). On the basis of coroner

Table 13. Cancer incidence and population exposure to electromagnetic fields

Reference	Subjects	Cancer deaths
Wertheimer & Leeper (1979)	Children living near high current configurations	Increased leukaemia
Fulton et al. (1980)	Children living near high current configurations	No increased leukaemia
Wertheimer & Leeper (1982)	Adults living near high current configurations	Increased cancer
Coleman et al. (1985)	Persons living near high voltage lines	No increased leukaemia
Myers et al. (1985)	Children living near high voltage lines	No increased cancer
Rodvall et al. (1985)	Persons living near high voltage lines	No increased cancer
Tomenius (1986)	Children living near high voltage lines	No increased leukaemia Increased nervous system tumours
McDowall (1986)	Persons living near high voltage lines	No increased cancer

and police records from various urban and rural regions within a 5000 km^2 area in the Midlands of England, a statistically significant increase in suicide rate was found among individuals who lived in residences where the 50-Hz field intensity exceeded 0.15 μT at the front entrance. A subsequent statistical analysis of the same data indicated that the cumulative probability ratio for the incidence of suicide increased above the null effect level of unity for residential 50-Hz magnetic field intensities exceeding 15 nT (Smith, 1982). However, oscillations occurred in the cumulative probability ratio as a function of increasing magnetic field intensity, and at 0.2 μT, the ratio for the "urban" study group was consistent with the absence of any 50-Hz magnetic field effect. From an epidemiological perspective, the lack of a clear-cut dependence of the suicide incidence on magnetic field intensity suggests that the apparent correlation between these variables may be purely fortuitous. An extension of the studies initiated by Perry et al. (1981), using a significantly larger population of individuals, will be required before any firm judgement can be made regarding the proposed correlation between suicide incidence and ELF magnetic field exposure. Thus, these data cannot serve as a basis for the evaluation of possible health effects, particularly as McDowall's (1986) data based on an analysis of mortality in a group of nearly 8000 persons, identified as living in the vicinity of electrical transmission facilities, did not support an association with suicide.

7.3 Studies on Human Volunteers

A number of research workers (Mantell, 1975; Hauf, 1976, 1982; Denisov et al., 1979; Sander et al., 1982; Kholodov & Berlin, 1984) have performed controlled studies on human volunteers in laboratories where the field strength and exposure duration were accurately known. The strongest fields and lengths of exposure were used by Sander et al. (1982) and Kholodov & Berlin (1984).

Sander et al. (1982) exposed human volunteers to 50-Hz magnetic fields of 5 mT. These exposures did not produce any effects with the exception of some minor variations in certain haematological parameters. All of the studies on human volunteers exposed to relatively weak magnetic fields produced negative results (Table 14).

Table 14. Effects of ELF magnetic fields on man

Exposure	Effect	Reference
0.3 mT, 50 Hz (for 3 h)	No effect on reaction time or EEG	Mantell (1975); Hauf (1976)
3 mT, 10 Hz	Threshold for perception or sensation	Denisov et al. (1979)
5 mT, 50 Hz (for 4 h/ day for 1 week)	No effect on many physiological parameters	Sander et al. (1982)
3 mT, 10 Hz	Threshold for perception or sensation	Kholodov & Berlin (1984)
2 - 10 mT, 15 - 20 Hz; 10 mT, 50/60 Hz	Threshold for perception of magnetophosphenes	Various authors (Table 7)
60 mT, 50 Hz	Threshold for visually evoked potentials	Silny (1986)

Kholodov & Berlin (1984) exposed the head, arms, and legs of human volunteers to determine the thresholds for sensation or perception of magnetic fields. They reported that, for pulsed magnetic fields (f = 10 Hz) the threshold was about 3 mT, for sweeping magnetic fields, about 0.5 mT, and for static fields, about 8 mT.

When the hands of human volunteers were exposed to static magnetic fields of up to 0.1 T for up to 30 min, skin temperature and sensitivity decreased, and capillary spasms were reported (Roschin, 1985).

8. HEALTH EFFECTS ASSESSMENT

The process of making a health risk evaluation is quite complex and involves consideration of such concepts as numerical values of risk, acceptability of risk, reasonable or comparative risk, public perception of risk, and cost-benefit analyses (Sinclair, 1981).

In making an assessment of the health risks from exposure to magnetic fields, criteria must be developed to identify which effects are to be considered a hazard for human health. The difficulty in defining the health hazard occurs when value judgements are involved that may not be based on scientific analysis.

Strict guidelines must be established prior to reviewing the literature on the biological effects of exposure to magnetic fields. Certain studies are conducted to identify underlying mechanisms of interaction. Many of these will be conducted on biological systems exposed in vitro to magnetic fields. Health effects assessments cannot be based on in vitro studies alone, because effects found in vitro may not necessarily occur in vivo. In vitro studies make it possible to determine the toxicity of an agent in increasingly complex steps. For example, effects on solutions of biological molecules might be used as a model system to study predominant mechanisms of action. Uncomplicated systems can assist in the exploration and evaluation of mechanisms and may serve as a useful basis for designing studies at the next level of biological complexity, the cellular level. By restricting the complexity of the experimental system, there will be less chance of possible subtle effects being masked by gross or dominant effects.

Thus, health agencies can place only limited value on in vitro studies. However, the in vitro results may indicate that a cautious or prudent approach should be adopted when setting standards. Once mechanisms of interaction are understood and found to occur in laboratory animals, the next step is to determine if it is possible to extrapolate the results to man.

Present knowledge of the interaction mechanisms operating when biological systems are exposed to magnetic fields is not sufficient to predict theoretically the whole range of effects of exposure to these fields, particularly the long-term effects. Thus, care must be applied in attempting to predict or extrapolate effects in man from effects found in laboratory animals.

An approach to making a health risk assessment is to evaluate the available data on exposure levels and bioeffects to determine if thresholds for effects occur (Repacholi,

1985b). It should be noted that, in undertaking such an evaluation, only reports that provide adequate information on experimental technique and dosimetry should be used. Ideally, only data that have been reproduced or substantiated by independent laboratories and have a direct bearing on health risk should be considered.

If possible, the health risk assessment should be based on well-conceived, -conducted, and -analysed epidemiological studies. Unfortunately, epidemiological studies on human beings exposed to magnetic fields tend to suffer from one or more of the following deficiencies: small numbers of subjects (resulting in low statistical accuracy); a lack of adequate dosimetry or ill-defined exposure conditions; lack of information on confounding variables, such as exposure to other physical or chemical agents; and a lack of a properly matched, stable control group that would provide unequivocal interpretation of the data to give a direct causal relationship with the hazardous physical agent.

Health risk analysis for the development of standards might adopt a phenomenological or conservative approach (Kossel, 1982; Repacholi, 1983a,b). In this case, it is assumed, until more information becomes available, that exposure to fields that produce an adverse biological effect could be hazardous, since later studies may reveal that the biological effect was a precursor to real injury.

8.1 Static Magnetic Fields

From the available data summarized in section 7, it can be concluded that short-term exposure to static magnetic fields of less than 2 T does not present a health hazard. Because of the lack of experimental data and from analysis of established mechanisms of interaction, exposure to fields above 2 T cannot yet be evaluated.

8.2 Time-Varying Magnetic Fields

In evaluating human exposure to time-varying magnetic frequencies up to about 300 Hz, it is possible to use an organ-dose concept (Bernhardt et al., 1986). This is based on two assumptions:

(a) There are no indications that a specific time-varying magnetic field effect exists at tissue field strengths below the value at which induced eddy currents may cause biological effects. Reports on calcium efflux (Adey 1981; Blackman et al., 1985b) and on effects in chick embryos (Delgado et al., 1981, 1982; Ubedą et al., 1983;

Juutilainen et al., 1986), if confirmed, would appear to be due to other mechanisms.

(b) When possible health risks for man from exposure to time-varying magnetic fields are evaluated, the biological effects mainly considered are those that originate from a direct action on the cells in nerve and muscle tissues. The physical quantity determining the biological effect is the induced electric field strength in the tissue surrounding the living cell.

There is a considerable amount of experimental data on stimulation thresholds for different nerves and muscle cells, often expressed in the form of electric current density values and not as field strength values. Only a few papers provide data on field strength thresholds. Therefore, the current density may be used as the decisive parameter in the assessment of the biological effects at the cellular level. Field strength and current density are related by the conductivity of the medium.

Selection of the current density as a measure of an action at the cellular level also makes it possible to extrapolate conditions in the human body from experimental animal studies or from measurements taken on isolated cells, by way of mutual comparison of the current densities. It seems irrelevant whether the electric current density surrounding a cell is introduced into the body through electrodes or induced in the body by external magnetic fields. However, the current paths within the body may be different in the two cases.

Several ranges of current densities may be considered.

(a) Up to 10 mA/m^2

It can be assumed that a current density of less than 1 mA/m^2, induced by an external magnetic field, should not produce adverse neurological or behavioural effects, since naturally flowing currents in the brain are of the same order of magnitude. Similar arguments pertain to fields that produce current densities of less than 10 mA/m^2 in the heart. In general, the endogenous current densities in major tissue and organ systems, other than the heart and brain, are below the 1 mA/m^2 level. Cellular responses in various tissues have been observed as shown in Fig. 8, and effects on tissue (bone) repair have been noted.

(b) 10 - 100 mA/m^2

In this range, electro- and magnetophosphenes are observed. Magnetophosphenes can be considered harmless for a

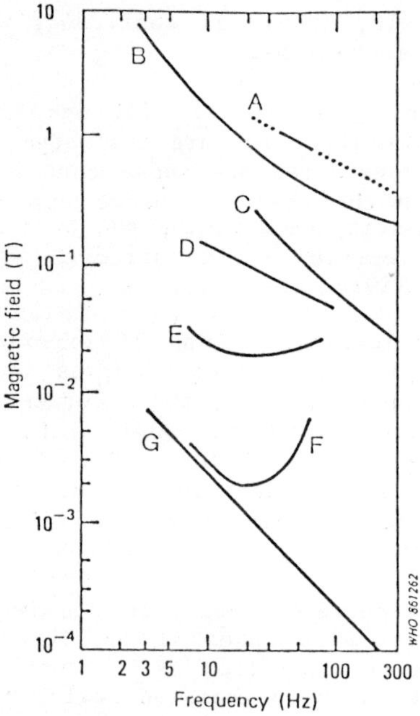

Fig. 8: Threshold values of the sinusoidal magnetic field for different biological effects.

Curve A: cardiac stimulation (fibrillation) threshold. From: Silny (1986).
Curve B: extra-systole stimulation threshold (with stimulus duration greater than 1 s).
Curve C: threshold curve for stimulating excitable cells and for producing membrane potential changes in single cells.
Curve D: effects on visual evoked potentials in man. From: Silny (1981).
Curve E: threshold for producing clearly distinguishable magnetophosphenes. From: Silny (1981).
Curve F: threshold for producing detectable phosphenes in about 10% of volunteers. From: Bernhardt (1985).
Curve G: current density curve of 1 mA/m² calculated for a 6- or 7.5-cm radius tissue loop with a conductivity of 0.2 S/m, serving as a model for the heart or head.

short exposure; however, the consequences of a long-term exposure with current densities at, or above, 10 mA/m² are not known. Furthermore, this current density will produce a membrane potential of the order of 0.1 mV (Bernhardt et al., 1986), which may influence the activity in other neurons. The results of electrophysiological studies have shown that information can be transferred between neuronal elements, even without action potentials (Schmitt et al., 1975). It must be expected that current densities that are below the nerve stimulation thresholds, may still influence brain function associated with electrical activity.

(c) 100 - 1000 mA/m^2

It has been shown by a number of research groups that current densities in this range, which result from electric currents applied in vitro and in vivo to mammalian central nervous tissue, can influence neuron excitability without causing direct stimulation. Much of this work has been carried out using rat hippocampal slices (Bawin et al., 1975, 1978, 1984, 1986) and guinea-pig hippocampal slices (Jeffries, 1981). The thresholds for stimulation of sensory receptors and of nerve and muscle cells may also lie in this range. It is possible that such stimulation could be hazardous. An unexpected stimulation of muscle tissue may lead to a dangerous reaction. Changes in excitability or the direct stimulation of central nervous tissue may lead to adverse changes in mental function.

(d) Above 1000 mA/m^2

An increased probability of ventricular fibrillation occurs at current densities above 1000 mA/m^2. The probability of this effect increases with both duration of exposure and current density magnitude. Continuous (tetanic) muscle contraction may also occur. In studies where 50/60-Hz electric currents have been applied to human volunteers via electrodes, tetany of the muscles concerned with breathing has been produced which, obviously, would be fatal if prolonged.

A summary of the ranges of induced currents that produce these possible effects is given in Table 15.

Table 15. Induced current density ranges between 3 and 300 Hz for producing biological effects

Current density (mA/m^2)	Effects
< 1	Absence of established effects
1 - 10	Minor biological effects reported
10 - 100	Well established effects, visual (magnetophosphenes) and possible nervous system effects; facilitation of bone fracture reunion reported
100 - 1000	Changes in central nervous system excitability established; stimulation thresholds; possible health hazards
> 1000	Extrasystoles, ventricular fibrillation possible; definite health hazards

In terms of a health risk assessment, it is difficult to correlate the internal tissue current densities with the external magnetic field strengths. Calculation of current densities using Faraday's law is complicated by the fact that the exact current paths depend in a complex way on the distribution and the conducting properties of the body tissues. Current densities induced in human beings and animals are extremely non-uniform. Current enhancements have been predicted in the human neck, axillae, and lower pelvic region for exposure to a horizontal ELF magnetic field (Kaune & Curley, 1986). There are differences in the conductivity of the white and grey cerebral matter. Furthermore, the effective diameter of the current pathways (loops) is not known. However, using "worst case" assumptions, an estimate of the order of magnitude for "safe" and dangerous magnetic field strengths and their frequency dependence can be made (Bernhardt, 1979, 1985).

The threshold field strengths and induced current densities required to produce visual effects by exposure to time-varying magnetic fields have been studied as a function of frequency (sections 4.2 and 6.1). In addition, the effects of electrical stimulation on cell membrane potentials, sensory receptors, and cardiac, nerve and muscle tissues have been characterized as a function of frequency (section 6.2). The frequency dependence of the thresholds for the direct electrical stimulation of cells and tissues, as well as the thresholds for magnetic field generation of phosphenes and for altering the VEP, have recently been summarized by Bernhardt (1985). By calculating the magnetic flux density that would produce current densities in tissues comparable with those produced by direct electrical stimulation, Bernhardt (1985, 1986) has constructed a family of curves representing the approximate threshold field levels necessary to produce electrical stimulation of cells and tissues by time-varying magnetic fields with a sinusoidal waveform. These threshold field levels are plotted in Fig. 8 as a function of frequency in the ELF range. Seven curves are shown in this figure, including some experimental data as explained in the caption.

With the possible exception of production of magnetophosphenes, over the entire ELF range, the threshold field levels that produce stimulating effects in various target organs and tissues are greater than those that induce a current density of 1 mA/m^2 in the brain or heart. This observation is consistent with the results of cell and tissue studies summarized in section 6.3, which indicate that the threshold current density for which perturbations are consistently observed is approximately 10 mA/m^2.

The values given in Fig. 8 for the current densities are applicable only to the peripheral regions of the heart or the head. For zones closer to the centre of the heart or the head (having a shorter current path), higher strengths of the magnetic field are necessary to induce the same current densities. From Fig. 8, a magnetic field strength that is considered not to produce any biological effect is about 0.4 mT for 50 or 60 Hz. Although some experimental data fit satisfactorily into Fig. 8, it must be understood that the figure only gives an idea of the magnitude of the current density in the body. Mean values were taken as the basis to determine the distribution of the electric field in the heart and the head, where the exact current paths are not known. Local increases in the internal field strength cannot be precluded. The extent of high local field strengths needs further elucidation by continued studies.

Safety factors may be defined more precisely only after further studies. This has to be considered in the case where curve G in Fig. 8 is used to evaluate human exposure to time-varying magnetic fields or to provide a basis for discussion on the definition and determination of personnel exposure limits.

8.3 Conclusions

1. Only a few mechanisms of the interaction of biological tissue with magnetic fields have been established. Some of the biological effects data suggest that other mechanisms may play a role, but these have yet to be confirmed experimentally. Thus, only a preliminary assessment of the human health risks from exposure to magnetic fields can be made.

2. A number of lower organisms have shown a remarkable sensitivity to the earth's magnetic field, because of highly developed receptors. Similar receptors have not been found in human beings.

3. For human exposure to static magnetic fields, it is not possible to make any definitive statement about the safety or hazard associated with short- or long-term exposure to fields above 2 T. Available knowledge suggests the absence of any measurable effect of static fields on many major developmental, behavioural, or physiological parameters in higher organisms. Recent medium-term (days) studies on exposure of animals to static fields of up to 2 T have not demonstrated any detrimental effects.

4. From the scientific data base on higher organisms exposed to magnetic fields, only 4 types of effect can be regarded as established. The first three may be explained by plausible

mechanisms of interaction and produce a basis for extrapolation to man. These effects are:

(a) induction of electrical potentials and magnetohydrodynamic effects within the circulatory system;

(b) the formation of magnetophosphenes with a time rate of change of magnetic field exceeding 0.3 T/s at 17 Hz; the effect depends strongly on frequency (compare Fig. 8);

(c) direct stimulation of nerve and muscle cells by very short (less than 1 ms) pulses of rapidly changing magnetic fields (several thousand T/s). Current densities are estimated to exceed 1000 mA/m^2. These effects are strongly frequency dependent and may exhibit lower thresholds (100 - 1000 mA/m^2) under more favourable stimulus conditions (10 - 100 Hz).

(d) other cellular and tissue alterations when the induced current densities exceed approximately 10 mA/m^2.

5. For human exposure to time-varying magnetic fields, it seems reasonable to assume that a health risk assessment can be made on the basis of significant perturbations of biological functions caused by electric currents induced by the fields. Available data suggest that, when current densities less than 10 mA/m^2 are induced in tissues and extracellular fluids, the induction of adverse health effects is unlikely. However, the possibility of some perturbing effects occurring following long-term exposure cannot be excluded.

The time-varying fields that induce currents in the body depend critically on the waveform and pulse shape. In this regard, the peak instantaneous current densities appear to be important. Furthermore, the frequency dependence of effects produced by time-varying fields has to be taken into consideration.

9. STANDARDS AND THEIR RATIONALES

With advances in technology resulting in increasing numbers of devices using magnetic fields, the potential for human exposure to these fields has increased to the point that valid questions are raised concerning safety.

Except for the USSR (USSR, 1970, 1978, 1985) and the Federal Republic of Germany (1986), no countries have developed, or are developing, mandatory standards limiting magnetic field exposure because, until recently, there was only a small probability of human exposure to magnetic fields strong enough to cause adverse health effects. However, with the advent of high-energy accelerators and fusion reactors using strong magnets, magnetic levitation systems for transport and, most recently, the application of magnetic resonance techniques in diagnostic medicine, serious consideration has been given to developing exposure limits in various countries.

A safety standard is a general term, incorporating both regulations and guidelines, and is defined to be a set of specifications or rules to promote the safety of an individual or group of people. A regulation is promulgated under a legal statute and is referred to as a mandatory standard. A guideline generally has no legal force and is issued for guidance only - a voluntary standard. Safety standards can specify maximum exposure limits and other safety rules for personnel exposures, or provide details on the performance, construction, design, or functioning of a device.

The purpose of this section is to briefly summarize the existing standards on magnetic fields and to discuss their scientific basis.

9.1 Static Magnetic Fields

Only a few guidelines limiting occupational exposure to static magnetic fields have been developed. The limits of human exposure to static magnetic fields in the USSR, US Department of Energy, and certain accelerator laboratories in the USA, and the CERN Accelerator Laboratory in Geneva are summarized in Table 16. Only one standard (USSR, 1978) has been promulgated to regulate static magnetic fields. A new DIN-VDE draft electromagnetic field standard is being discussed in the Federal Republic of Germany (1986) and this includes 0 Hz magnetic fields.

The earliest static magnetic field guidelines were developed as an unofficial recommendation in the USSR (Vyalov, 1967). Clinical investigations (Vyalov et al., 1964; Vyalov & Lisichkina, 1966; Vyalov, 1971, 1974) formed the basis for the Soviet Standard (USSR, 1978). The standard requires that the

Table 16. Limits of occupational exposure to static magnetic fields

Author	Field	Exposure time	Body region	Comments
USSR (1978)	0.01 T	8 h	whole body	regulation issued by Ministry of Health
Stanford Linear Accelerator Center (1970)	0.02 T	extended (h)	whole body	unofficial, occupational
	0.2 T	short (min)	whole body	
	0.2 T	extended (h)	arms, hands	
	2 T	short (min)	arms, hands	
US Department of Energy (DOE) (Alpen, 1979)	0.01 T	8 h	whole body	recommended to DOE contractors
	0.1 T	1 h or less	whole body	
	0.5 T	10 min or less	whole body	
	0.1 T	8 h	arms, hands	
	1 T	1 h or less	arms, hands	
	2 T	10 min or less	arms, hands	
CERN Accelerator Lab, Geneva (NRPB, 1981)	0.2 T	minutes	whole body	Recommended practice
	2 T	short	hands, arms and feet	
Lawrence Livermore National Laboratory (LLNL, 1985)	0.06 T	day	trunk	maximum average/day in peak fields > 0.5 T
	0.06 T	day	trunk	maximum average/week in peak fields < 0.5 T
	0.6 T	day	extremities	maximum average/week (in peak fields < 0.5 T) or per day (in peak fields > 0.5 T)
	2 T	short (min)	whole body	peak exposure limit

static magnetic field strength at the work-place does not exceed 8 kA/m (0.01 T).

Three sets of guidelines recommending limits of occupational exposure to static magnetic fields exist in the USA. Two of these are applicable in high-energy physics laboratories, and the other is a US Department of Energy (DOE) guideline.

At the Stanford Linear Accelerator Center in California, unofficial guidelines were established in 1970. They suggest that the whole body or head of workers should not be exposed to static magnetic fields exceeding 0.02 T for extended periods (h) or fields exceeding 0.2 T to the arms and hands. For short periods (min), the whole body or head, and arms and hands should not be exposed to fields exceeding 0.2 and 2 T, respectively. The 2-T limit also allows film changes at Stanford's bubble chamber.

The Lawrence Livermore National Laboratory (LLNL, 1985) has drafted a set of policy guidelines for working in magnetic fields associated with the high-energy accelerators. The guidelines (Table 16) state that:

> Maximum exposure: Workers must never be exposed to fields exceeding 2 T, regardless of the duration of the exposure or the exposed part of the body;
>
> Fields less than 500 mT: If the peak field to which workers are exposed is less than 500 mT, personnel may be exposed to a week-long maximum average field strength of no more than 60 mT (measured at the torso) or 600 mT (measured at the extremities);
>
> Fields greater than 500 mT: If the peak exposure is greater than 500 mT, workers should be exposed to a daily maximum average field strength of no more than 60 mT (measured at the torso) or 600 mT (measured at the extremities).

In addition, the following restrictions are made: Always use caution signs indicating the presence of a magnetic field, whenever the field strength is 1 mT or greater. Use additional administrative controls or barricades (ropes or fences), whenever practical. Do not allow workers with cardiac pacemakers or other medical electronic implants into areas where the magnetic-field intensity exceeds 1 mT. Magnetic fields greater than this level can trigger a change in the operating mode of some pacemakers. Persons with small metallic implants (such as aneurysm clips) must also be stopped from entering an area where the field intensity is greater than 1 mT. Stronger magnetic fields may rotate or even remove aneurysm clips from the arteries to which they are attached. Workers with large metallic implants, such as hip prostheses, should be advised to avoid working anywhere inside the perimeter of 1-mT field intensity.

A rationale supporting the guidelines accompanies the document (LLNL, 1985). The 60-mT limit is set to 1 mV of the magnetohydnamic voltage (voltage generated by blood, an ionized fluid, moving in a fixed magnetic field) in an obese person engaged in moderately heavy work (cardiac output 10 litres/min). 2 T limits the rise in blood pressure to 1%.

The US Department of Energy (DOE) formed an ad hoc committee to review technologies that use magnetic fields, to make an assessment of the scientific literature on biological effects, and to establish guidelines for static magnetic fields, field gradients, and time-varying magnetic fields. In July 1979, the Alpen Committee (Alpen, 1979) made its recommendations to DOE as shown in Table 16. The guideline in some cases is a factor of 2 lower than that for continuous exposure at the Stanford Linear Accelerator Center. This guideline was recommended by the Department of Energy to its contractor organizations as an interim measure, until official standards are promulgated. Although the Alpen Committee made a review of the literature it has not published a rationale supporting the values recommended in their guideline. According to Tenforde, the 0.01 T limit was recommended for continuous exposure, because this represented the accepted threshold for magnetophosphene production by ELF magnetic fields, and the threshold for inducing measurable electrical potentials in the central circulatory system during exposure to static magnetic fields.

A similarly recommended practice for limiting static magnetic field exposures of workers exists at the CERN accelerator laboratory in Geneva (NRPB, 1981). CERN recommends that exposure of the hands, arms, and feet should not exceed 2 T for periods of the order of minutes. This is reduced by a factor of 10, if the head or whole body is exposed.

With the advent of magnetic resonance imaging (MRI), the need for occupational exposure limits has become more apparent, and other organizations that traditionally recommend occupational exposure limits have begun to address this need, despite a reluctance in the past to recommend limits for magnetic fields (Sliney, 1986).

9.2 Time-Varying Magnetic Fields

Except for guidelines limiting patient and operator exposure during clinical magnetic resonance imaging, the only standard limiting exposure of time-varying magnetic fields in the ELF range is the Soviet Standard (USSR, 1985), as shown in Table 17. The 50-Hz magnetic field standard (USSR, 1985) issued by the Ministry of Public Health of the USSR in January 1985 makes a distinction between continuous and pulsed fields and limits the duration of exposure, depending on the pulse characteristics. The limits for exposure to continuous wave fields equate to 7.5 mT for 1 h and 1.8 mT for 8 h. This standard seems to have been developed for arc welding, since pulsed field exposure occurs most frequently in welding. The scientific basis for this standard does not appear to have been published.

The Federal Republic of Germany (1986) is discussing extention of its current electromagnetic field standard (Federal Republic of Germany, 1984) down to 0 Hz.

9.3 Magnetic Resonance Imaging Guidelines

Comparing the magnetic field limits in Table 16 with the strength of the magnets used in MRI, it is not surprising that regulatory and health agencies have begun to look more seriously at this imaging modality (Repacholi, 1986). Hundreds of MRI machines have been installed throughout the world and concern about their safety has been expressed (Bore, 1985). Some of these machines use superconductive magnets with fields for diagnostic application up to about 2.0 T, and there are prototypes with magnets giving fields of 4 - 5 T. These prototypes are being studied to determine the feasibility of *in vivo* spectroscopy.

During the imaging procedure, lasting up to tens of minutes, the patient lies on a table and all parts of the body are exposed to strong static magnetic fields, changing (or time-varying) magnetic fields and radiofrequency radiation. Rapidly switched gradient fields are superimposed on the static field to allow spatial information to be obtained. These time-varying fields induce electric currents in the body.

Table 18 shows the guidelines on static and time-varying magnetic field exposure for the clinical examination of patients during MRI, recommended by the Center for Devices and Radiological Health (CDRH, 1982) of the US Department of Health and Human Services, the National Radiological Protection Board (NRPB, 1984) in the United Kingdom, the Federal Health Office (FHO, 1984) of the Federal Republic of Germany, and Health and Welfare Canada (Health and Welfare Canada, 1986).

In January 1984, the Health Council of the Netherlands (HCN, 1984) issued interim advice on the use of magnetic resonance imaging. This general document contains a section on possible health risks including statements such as: there are no risks to health from static magnetic field exposures up

Table 17. Maximum permissible levels of magnetic fields with a frequency of 50 Hz[a]

Duration of exposure (h)	Magnetic field strength A/m		
	Continuous and pulsed magnetic fields with pulse width $t_W \geq 0.02$ s and pause $t_p \leq 2$ s	Pulsed magnetic field $60 \text{ s} \geq t_W \geq 1 \text{ s}$, $t_p > 2$ s	Pulsed magnetic field $0.02 \text{ s} \leq t_W < 1 \text{ s}$, $t_p > 2$ s
1	6000	8000	10 000
1.5	5500	7500	9500
2	4900	6900	8900
2.5	4500	6500	8500
3	4000	6000	8000
3.5	3600	5600	7600
4	3200	5200	7200
4.5	2900	4900	6900
5	2500	4500	6500
5.5	2300	4300	6300
6	2000	4000	6000
6.5	1800	3800	5800
7	1600	3600	5600
7.5	1500	3500	5500
8	1400	3400	5400

[a] From: USSR (1985).

Note: The above regimes of pulsed exposures are used in welding.
t_W is the pulse width duration.
t_p is the pulse pause duration.
Magnetic flux density in mT = Magnetic field strength in A/m x $\frac{1.256}{10^3}$

to 0.5 T, and even exposures to fields up to 2 T appear safe. More research is needed to determine the safety of fields stronger than 2 T. The document also recommends that limits for time-varying magnetic fields and radiofrequency fields accepted in the USA (CDRH, 1982) or the United Kingdom (NRPB, 1981) should be followed.

Table 18. Guidelines on magnetic field exposure in clinical MR

Country	Static fields	Time-varying fields
USA CDRH (1982)	Patient - 2 T whole and partial body exposure	Patient - 3 T/s whole and partial body exposure
	Exposure exceeding these limits should be evaluated on an individual basis	
United Kingdom NRPB (1984)	Operator - 0.02 T (long periods, whole body); 0.2 T (long periods, arms, hands); 0.2 T (15 min, whole body) 2 T (15 min, arms, hands) Patient and volunteers - 2.5 T (whole and partial body exposure)	Patient and volunteers - 20 T/s (rms) periods of magnetic field change $\geq$ 10 ms or $(dB/dt)^2t < 4$ (rms) for duration of magnetic field change < 10 ms where dB/dt in T/s and t in s
Germany, Federal Republic of FHO (1984)	Patient - 2 T (whole and partial body exposure)	Patient - whole and partial body exposure: maximum induced current density 30 mA/m² or 0.3 V/m electric field strength for duration of magnetic field change of 10 ms or longer or (300/t) mA/m² or (3/t) V/m for duration of magnetic field change (t) shorter than 10 ms (t in ms)
Canada Health and Welfare Canada (1986)	Operator - 0.01 T (whole body during working day) - > 0.01 T (keep to minimum) Patient - 2 T (whole and partial body exposure)	Patient - 3 T/s (rms)

9.3.1 United Kingdom

The NRPB (1984) recommends that the following conditions should be fulfilled during the operation of magnetic resonance imaging equipment in the United Kingdom.

(a) Static fields

For people (patients and volunteers) exposed to the imaging process, the static magnetic field should not exceed 2.5 T for the whole or a substantial portion of the body. The NRPB Advisory Group formulating the guidelines suggested that static fields have been shown to affect certain chemical reactions *in vitro* and that reproducible changes in primate behaviour have been found in fields of several tesla. Although flow potentials are generated across blood vessels by the flow of blood perpendicular to the field, their biological significance at fields of a few tesla remains unclear. However, at 2.5 T, the peak flow potential is calculated to be approaching the depolarization threshold for myocardial muscle. Although only a fraction of this potential occurs across each cell, it was considered prudent to limit acute exposure to 2.5 T, until further information becomes available.

Occupational static field exposure limits are recommended for staff operating MRI equipment. Exposure for prolonged periods to more than 0.02 T for the whole body or 0.2 T for the arms or hands should be avoided. NRPB (1984) recommends that these limits may be increased to 0.2 T for the whole body and 2 T for the arms and hands for periods totalling less than 15 min at a time, provided intervals of about 1 h occur between such exposures.

These operator limits are essentially the same as those recommended by the Stanford Linear Accelerator Center (Table 16), where no adverse symptoms have been reported from staff working at the facility, since the introduction of their guidelines in 1970.

(b) Time-varying fields

For the time-varying fields, excluding radiofrequency fields, the NRPB (1984) recommends limits based on the duration of magnetic flux density changes (i.e., the time during which electric currents are being induced). When the duration of exposure exceeds 10 ms, exposures should not exceed root mean square (rms) rates of change of magnetic flux density (dB/dt) of 20 T/s for all persons (patients exposed to the imaging process, volunteers). For durations of change of less than 10 ms, the relationship $(dB/dt)^2t$ less than 4 should be observed where dB/dt is in T/s, and t is the duration of the change of the magnetic field in seconds. For continuously varying magnetic fields, such as sinusoidal fields, the duration of the change can be considered as half the period of the waveform.

The rationale for the NRPB guidelines is given in a publication by Saunders & Smith (1984). The NRPB Advisory

Group recognized that rapidly changing magnetic fields can induce electric currents in tissues that could be sufficiently large to interfere with the normal functioning of nerve cells and muscle fibres. These conduct electrical impulses in the form of localized membrane depolarization produced by the flow of ions and, above a certain threshold, give rise to sensation or muscle contractions. From experimental data it was inferred that the threshold would be lowest when the current pulse width (or duration of magnetic flux density change) exceeded about 10 ms.

It was felt that, although the sensation of magnetic phosphenes occurred at a threshold in man of about 1.3 T/s (at 20 Hz), this sensation of light flashes in the eye has not been shown to be hazardous. However, excitation of nerves and muscles could be hazardous, but requires exposure to high rates of change of magnetic flux density. The threshold for excitation depends on the pulse length and pulse repetition frequency of the induced current. Since insufficient information is available to define safe limits, they must be derived from effects of electric currents applied by electrodes. The threshold current density to induce ventricular fibrillation is 3 A/m^2. Thus, to achieve a factor of 10 safety margin, it was decided that MRI operating conditions should be such as to induce current densities that did not exceed 0.3 A/m^2 for a duration of magnetic flux density change greater than 10 ms.

For durations of the current pulse of half period (t) of less than 10 ms, the evidence suggests that, when t decreases, the threshold rms current density for inducing ventricular fibrillation increases. Experimental data suggest that the square of the rms current density multiplied by the duration (t) remains constant. The magnetic field vector in most MRI equipment is parallel to the longitudinal axis of the body (z-axis). The current density induced by time variation of the z-gradient is proportional to the conductivity, the inductive loop radius, and the rate of change of the magnetic field. Assuming the average value for tissue conductivity to be 0.2 S/m, the radius of the body to be 0.15 m, the limit applied to the rms current density of 0.3 A/m^2 for pulses or half periods of induced current exceeding 10 ms restricts the rms rate of change of the z gradient magnetic flux density to 20 T/s, when the duration of magnetic field change exceeds 10 ms. For durations shorter than 10 ms, (Saunders & Smith, 1984), the relationship for determining the limit for the time-varying field can be derived:

$$(dB/dt)^2 t < 4,$$

where dB/dt is in T/s, and t is in seconds

It was assumed that the current densities induced in the body by variation of the anterior-posterior (y) and lateral (x) gradients would not be significantly greater than for the z gradients.

(c) Other guidelines

(i) RF exposure of the patient and staff must be restricted so that the rise in temperature does not exceed 1 °C, as shown by skin and rectal temperature, or more than 1 °C in any mass of tissue not exceeding 1 g in the body.

(ii) Patients should be exposed only with the approval of a registered medical practitioner or research ethics committee.

(iii) Patients must be fully informed of the procedure and consent freely to it.

(iv) Only medically assessed suitable volunteers should be used in trials.

(v) Frequently exposed volunteers should have regular ECG checks.

(vi) It is prudent to exclude women in the first three months of pregnancy.

(vii) Special care is needed for patients with cardiac pacemakers or large metallic implants.

(viii) Warning notices should be posted indicating that magnetic and RF fields may affect pacemakers and electronic equipment.

9.3.2 USA

The recommendations issued by the Center for Devices and Radiological Health (CDRH, 1982) in the USA are intended to assist the medical profession and manufacturers in making health risk benefit assessments. Based on information available in the literature, it was suggested that, in the case of diagnostic magnetic resonance applications involving exposure to static magnetic fields not exceeding 2 T or time-varying fields not exceeding 3 T/s, the benefits outweigh the risks, within the current medical indications and contra-indications. Pregnant women should not be exposed as the safety of such exposure has not been established. It should be noted that the CDRH guidelines are not limits for patient exposure in MRI imaging investigations. The recommendations are essentially criteria that provide a demarcation between devices exceeding the magnetic field levels stated in the guidelines and therefore requiring further evaluation to determine if any health risk exists for the patient, and devices operating below the levels given in the guidelines.

The recommendations for the magnetic field levels were determined after consideration of existing unofficial standards and recommendations and their rationales, and a review of the scientific literature. The scientific rationale for the guidelines is essentially that proposed by Budinger (1981). Budinger concluded, after a review of the bioeffects literature and a theoretical analysis of the known interaction mechanisms of static magnetic fields with biological systems, that harmful effects on human beings or reproducible cellular, biochemical, and genetic effects have not yet been observed and are not expected at fields of less than 2 T. For changing magnetic fields, Budinger concluded that the thresholds for effects of induced currents is above that produced by 1 - 100 Hz sinusoidal fields of strength 5 mT. However, he did note that potential biological effects due to differences in waveform, repetition rate, peak magnetic field, and duration of exposure required further study.

The CDRH also recommends that the radiofrequency field exposure of the patient should be limited, so that the specific absorption rate (SAR) does not exceed 0.4 W/kg, averaged over the whole body, or 2 W/kg, averaged over any gram of tissue.

9.3.3 Federal Republic of Germany

The Federal Health Office (FHO, 1984) has made recommendations to physicians who work with clinical MRI devices. It is stated that no adverse health effects on patients, operators, or any other persons in the vicinity of MRI equipment have been detected so far. However, possible effects on the body can be estimated from induced currents and potentials in the body. The guidelines for static and time-varying magnetic fields are based on these estimations and study of the literature. It is stated that, if there is compliance with these recommendations, any detrimental effects will be detected at the earliest possible time. A translation of the original guidelines from German to English is provided in Bernhardt & Kossel (1985).

(a) Static fields

The FHO recommends that patients imaged in an MRI facility should not be exposed to static magnetic fields exceeding 2 T. If patients are exposed to fields higher than 2 T, they should be monitored for cardiac and circulatory function.

The rationale for determining this value is as follows: orientation effects are observed in such systems as DNA, retinal rods, and sickle cells at static field strengths above 1 - 2 T. Electric potentials induced in flowing blood exposed to static fields above 0.3 T have been noted in ECG measurements in animals, but no adverse health effects have been observed in animals exposed to fields up to 10 T. However, the potential differences induced by cardiac contractions in a magnetic field exceeding 2T may impair the excitation stimulation or conduction of excitation.

(b) Time-varying fields

Time-varying magnetic fields induce electrical potentials, the size of which depends on the magnetic field strength, pulse duration and frequency. Using essentially the reasoning outlined by Bernhardt (1985) for estimating the values of induced electric potentials and currents that are likely to cause biological effects, the FHO (1984) recommends that patients should not be exposed to time-varying magnetic fields having a duration of magnetic field change equal to or greater than 10 ms, which induce electric fields greater than 0.3 V/m or current densities exceeding 30 mA/m^2. If the duration of the magnetic field change of the time-varying fields is less than 10 ms, then the maximum induced electric field is 3/t V/m and maximum induced current density is 300/t mA/m^2 (0.3/t A/m^2), where t is the duration of magnetic field change in milliseconds. The MRI machine manufacturer must inform the purchaser of the operating conditions that will result in the induced field strength and current density remaining below the recommended values. If these values are capable of being exceeded by the machine, the manufacturer must prove that it is safe.

A brief rationale for these recommended values is given in FHO (1984). Compared with natural currents, induced current densities of 1 mA/m^2 have no detectable effect on the body. Current densities of 10 mA/m^2 induce effects that depend on the frequency of the time-varying magnetic field, but do not pose hazards. At frequencies between 10 and 50 Hz, magnetic fields above 5 mT produce magnetophosphenes. Ventricular fibrillation may be caused if the magnetic field induces current densities exceeding 1000 mA/m^2 or electric fields exceeding 100 mV/cm.

The final values of induced electric field and current density represent estimates, based on studies and theoretical calculations (described by Bernhardt, 1985), that are thought to provide a wide margin of safety.

(c) Other recommendations

1. Exposure to radiofrequency fields should be such that the SAR does not exceed 1 W/kg (whole body) or 5 W/kg (partial body - per kg of tissue, except the eyes).

2. Prior to patient examination, care must be taken with regard to implants made of ferromagnetic materials, implanted cardiac pacemakers, dislocation of catheters, vascular clips, and the like.

3. Patients must undergo additional medical examinations (described in FHO, 1984), if the recommended exposure limits are exceeded.

9.3.4 Canada

The Bureau of Radiation and Medical Devices of Health and Welfare Canada (Health and Welfare Canada, 1986) has published a safety code containing guidelines on exposure to electromagnetic fields from magnetic resonance clinical systems. The document contains information on levels of exposure for typical devices, exposure guidelines from various countries, a summary of health effects from magnetic and radiofrequency fields, and guidance on exposure of patients and operators. Details of the guidelines are given in Table 18.

Health and Welfare Canada (1985) have also published recommendations to ensure the protection of patients and operational personnel from potential hazards in MRI. This report contains recommendations on magnetic fields as shown below:

(a) static magnetic fields must be below 0.5 mT in unrestricted areas;

(b) entrance to areas in excess of 1.5 mT must be strictly controlled, to prevent introduction of magnetic material by patients, operational personnel, and visitors;

(c) equipment for cardiopulmonary resuscitation must be available and usable within the imaging room and, if possible, in areas where the field exceeds 10 mT;

(d) static magnetic fields should not exceed 2.5 T (this differs from the 2 T recommended in Health and Welfare Canada (1986)); and

(e) time-varying magnetic fields should not exceed 3 T/s.

10. PROTECTIVE MEASURES AND ANCILLARY HAZARDS

Protective measures for the industrial and scientific use of magnetic fields can be categorized as engineering design measures, the use of separation distance, and administrative controls. Another general category of hazard control measures, namely personal protective equipment (e.g., special garments and face masks) do not exist for magnetic fields. However, protective measures against ancillary hazards from magnetic interference with emergency or medical electronic equipment and for surgical and dental implants are a special area of concern regarding health aspects of magnetic fields. The mechanical forces imparted to ferromagnetic implants and loose objects in high-field facilities require that precautions be taken.

The techniques to minimize needless exposure to high intensity magnetic fields around large research facilities generally fall into three types:

(a) Distance and time

Limit human access and/or occupancy duration in locations where field strengths are high. Since the external magnetic flux density decreases with distance from the source, separation distance is a fundamental protective measure. For example, at large distances from a static magnetic field dipole source, the field decreases approximately as the reciprocal cube of the separation distance.

(b) Magnetic shielding

The use of ferromagnetic core materials restricts the spatial extent of external flux lines of a magnetic device. External enclosures of ferromagnetic materials can also "capture" flux lines and reduce external flux densities. However, shielding is normally an expensive control measure and of limited use for scientific instruments. Furthermore, it has not generally been shown to be cost-effective for large installations when compared to the use of separation distance (Hassenzahl et al., 1978).

(c) Electromagnetic interference (EMI) and cardiac pacemakers

Certain types of modern cardiac pacemakers exhibit malfunction in response to EMI produced either by endogenous myopotentials or by external sources such as high-voltage systems. The modern implantable pacemakers are microprocessor-controlled and function in a "demand" mode in which stimulatory pulses are delivered to the heart, only if it fails to exhibit intrinsic electrical activity. The endogenous cardiac activity is detected by a signal-sensing circuit, in order to avoid competitive pacing between the pacemaker's stimuli and the heart's intrinsic activity. The modern pacemakers also contain a noise detection circuit that can discriminate electric fields with different frequencies and waveforms from those associated with the heart's bioelectrical activity. When EMI is sensed, the demand pacemaker reverts to a fixed-rate pacing mode, which may be asynchronous with the normal cardiac activity. This pacing mode is frequently referred to as the "reversion" or "noise" mode of operation, and can be undesirable if the pacemaker signals are competitive with the intrinsic cardiac electrical activity.

Two different configurations of electrode leads are used in pacemakers, and these have very different sensitivities to EMI. In one type, termed the "bipolar" design, both leads are implanted within the heart at a typical separation distance of 3 cm. In the second type, termed the "unipolar" design, the cathode lead is implanted in the heart and the pacemaker case serves as the anode. Because of the considerable physical separation of the anode and cathode leads in the unipolar design, this type of pacemaker provides a large antenna for the reception of EMI. Of the two designs for pacemaker electrode configurations, only the unipolar type has been found to be sensitive to EMI. Among the 350 000 - 500 000 individuals in the USA who have implanted pacemakers, approximately 50% have models with the unipolar electrode design.

During the past decade, several laboratory tests and studies on pacemaker patients have been conducted to assess the response of different pacemaker models to power-frequency electric and magnetic fields (Jenkins & Woody, 1978; Butrous et al., 1983; Griffin, 1985; Moss & Carstensen, 1985). Two types of pacemaker malfunction have been observed in response to EMI: (a) an aberrant pacing rate, with irregular or slow pacing; and (b) reversion to fixed-rate (asynchronous) pacing. The probability that a malfunction will occur in the presence of an external electromagnetic field is strongly dependent on the pacemaker model, since some manufacturers have incorporated a feature into their pacemaker models that automatically decreases the sensitivity of the amplifier circuit when EMI is sensed. These specific brands of pacemaker thereby avoid reversion to an asynchronous mode in response to EMI.

Griffin (1985) estimated the total population of pacemaker patients in the USA who might be at serious risk from the effects of EMI. He assumed that: (a) 350 000 - 500 000 individuals wear pacemakers; (b) 50% of the pacemakers are of the unipolar design; (c) 10 - 20% of the unipolar pacemakers are highly sensitive to EMI; and (d) 20 - 25% of the patients with sensitive pacemakers are totally dependent on the pacemaker to sustain their cardiac rhythm. With these assumptions, it can be calculated that approximately 3500 - 12 000 pacemaker wearers might be at serious risk from EMI. However, it must be borne in mind that only a small fraction of the individuals at risk are likely to encounter a source of EMI during the time periods when their cardiac function is totally dependent on an implanted pacemaker. The above estimate of the population at risk must therefore be regarded as an upper limit that perhaps greatly overestimates the actual probability of the occurrence of a potentially fatal pacemaker malfunction in response to EMI.

Both power-frequency electric and magnetic fields have been found to introduce EMI that can alter the functioning of many commercially available pacemakers. In studies on 11 patients with 7 different implanted pacemaker models from 4 manufacturers, Moss & Carstensen (1985) observed alterations in pacemaker function during exposure to 60-Hz electric fields ranging from 2 - 9 kV/m. Only models produced by 2 out of the 4 manufacturers were sensitive to EMI from fields of this strength. A similar set of observations was made by Butrous et al. (1983).

A total of 26 pacemaker models were examined by Jenkins & Woody (1978). Twenty of these units were found to revert to an asynchronous mode of pacing or to exhibit abnormal pacing characteristics in 60-Hz fields ranging from 0.11 to 0.4 mT, with the average threshold field level for an effect being 0.21 mT. The minimum value of dB/dt producing an effect was therefore 41.5 mT/s (for the 60-Hz, 0.11-mT field). Pacemaker malfunctions produced by power-frequency magnetic fields require field levels that are greater than those associated with high-voltage transmission lines and most other types of electrical systems. However, the fields in the immediate vicinity of various types of industrial machinery and appliances (section 3) are sufficiently strong to represent a potential source of EMI that could alter pacemaker functioning.

Pavlicek et al. (1983) found that a rapidly switched gradient field used in magnetic resonance imaging with a time variation of 3 T/s could induce potentials up to 20 mV in the loop formed by the electrode lead and the case of a unipolar pacemaker. This signal amplitude is sufficiently large to avoid rejection by the pacemaker's EMI discrimination circuitry, and could therefore be recognized as a valid cardiac signal.

Pacemaker malfunctions can also be caused by static magnetic fields, which produce closure of a reed relay switch used to test the pacemaker's performance while operating in a fixed rate pacing mode. On the basis of a study of pacemakers produced by 6 major manufacturers, Pavlicek et al. (1983) found the most sensitive model to exhibit reed switch closure and reversion to fixed-rate pacing in a 1.7-mT static field. Field levels of 1.7 - 4.7 mT were observed to produce closure of the reed switches in all of the models tested. All of the models were also found to experience forces and torques when placed in MRI devices operated at fields of up to 0.5 T. Two of the pacemakers experienced a torque that was strong enough to produce significant movement of these units within tissue.

(d) Administrative measures

The use of warning signs, and special access areas to limit exposure of personnel near large magnet facilities has been of greatest use to control exposure. Administrative controls, such as these, are generally preferable to using magnetic shielding, which can be extremely expensive. In some circumstances, for example MRI facilities, a combination of shielding, restricted access, and the use of metal detectors may be appropriate to avoid detrimental effects from exposure to high strength magnetic fields. Loose ferromagnetic and paramagnetic objects can be converted into dangerous missiles when subjected to intense magnetic field gradients. Avoidance of this hazard can only be achieved by removing loose metallic objects from the area and personnel. Such items as scissors, nail files, screwdrivers, and scalpels should be banned from the immediate vicinity.

Of particular concern in MRI, are the forces exerted by static magnetic fields on implanted metal objects such as aneurysm clips and pacemakers. Even the most modern pacemakers will malfunction when placed in an MRI machine (Erlebacher et al., 1986). New et al. (1983) also measured the magnetic torques exerted on 21 types of haemostatic clips and various other materials such as dental amalgam. Of the 21 clips, 19 of which were aneurysm clips, 16 showed a deflection near the portals of two magnets operating at 0.147 T and 1.44 T, respectively. Of the remaining materials tested, only a shunt connector demonstrated significant ferromagnetic properties. The non-magnetic materials were primarily composed of austenitic stainless steel. Surgical clips composed of tantalum or titanium are also non-ferromagnetic. Clips composed of martensitic stainless steels are ferromagnetic and experience significant forces and torques in static magnetic fields. These findings indicate a clear requirement for strict administrative controls in determining whether patients bearing medical implants could be adversely affected by the fields present in MRI devices.

REFERENCES

AARHOLT, E., FLINN, E.A., & SMITH, C.W. (1981) Effects of low-frequency magnetic fields on bacterial growth rate. Phys. biol. Med., 26: 613-621.

AARHOLT, E., FLINN, E.A., & SMITH, C.W. (1982) Magnetic fields affects the Lac operon system. Phys. biol. Med., 27: 606-610.

ABASHIN, V.M. & YEVTUSHENKO, G.I. (1984) The mechanisms of biological effects of low frequency electromagnetic fields in the biosphere. In: Biological effects of electromagnetic fields, Moscow, Nauka, Vol. 2, pp. 193-201.

ABRAMOVICH-POLJAKOV, D.K., KLEINER, A.I., KOLODUB, F.A., KRAKOVSKAYA, S.P., NEDBAILS, E.P., PENOVA, V.N., SMOLJANOVA, N.S., TRIKOZA, V.A., & STEINGERZ, L.A. (1979) [Clinical characteristics of effects of electromagnetic fields from contact welding.] Vrach. Delo, 4: 106-119 (in Russian).

ACETO, H., Jr, TOBIAS, C.A., & SILVER, I.L. (1970) Some studies on the biological effects of magnetic fields. IEEE Trans. Mag., 6: 368-373.

ADEY, W.R. (1980) Frequency and power windowing in tissue interactions with weak electromagnetic fields. Proc. IEEE, 68: 119-125.

ADEY, W.R. (1981) Tissue interactions with non-ionizing electromagnetic fields. Physiol. Rev., 61: 435-514.

ADEY, W.R. (1983) Some fundamental aspects of biological effects of extremely low frequency (ELF). In: Grandolfo, M., Michaelson, S.M., & Rindi, A., ed. Biological effects and dosimetry of non-ionizing radiation, New York, London, Plenum Press, 561-580.

AKOYUNOGLOU, G. (1964) Effect of a magnetic field on carboxydismutase. Nature (Lond.), 202: 452-454.

ALEKSANDROV, M.S., BAKLENEVA, A.M., GLADSTEIN, N.D., OZEROV, V.P., POTAPOV, A.V., & REMIZOV, L.T. (1972) [Fluctuations of the earth's electromagnetic field,] Moscow, Nauka, p. 195 (in Russian).

ALPEN, E.L. (1979) Magnetic field exposure guidelines. In: Tenforde, T.S., ed. Magnetic field effects on biological systems, New York, London, Plenum Press, pp. 25-32.

AMER, N.M. (1965) The effects of homogeneous magnetic fields, ambient gas composition, and temperature on development of Tribolium confusum, Berkeley, University of California, Lawrence Radiation Laboratory (Report No. UCRL-16854).

AMERICAN INSTITUTE OF BIOLOGICAL SCIENCES (1985) Biological and human health effects of extremely low frequency electromagnetic fields, Arlington, Virginia, American Institute of Biological Sciences, p. 290.

AMINEV, G.A., AMINEVA, R.I., & SITKIN, M.I. (1967) [The influence of magnetic fields on the learning process of mice in a T-shaped maze.] In: [Problems of bionics,] Moscow, Nauka, pp. 338-340 (in Russian).

ANDRIANOVA, L.A. & SMIRNOVA, N.P. (1977) [Motor activity of muscles in a magnetic field of varying intensity.] Kosm. biol. aviakosm. Med., 11: 54-58 (in Russian).

ANIOLCZYK, R. (1981) Measurements and hygienic evaluation of electromagnetic fields in the environment of diathermy, welders, and induction heaters. Med. Pracy, 32: 119-128.

ANTONOWICZ, K. (1974) Possible superconductivity at room temperature. Nature (Lond.), 247: 358-360.

ARCHER, C.W. & RATCLIFFE, N.A. (1983) The effects of pulsed magnetic fields on chick embryo cartilaginous skeletal rudiments in vitro. J. exp. Zool., 225: 243-256.

ARNOLD, W., STEEL, R., & MUELLER, H. (1958) On the magnetic asymmetry of muscle fibres. Proc. Natl Acad. Sci. (USA), 44: 1-11.

ASANOVA, T.P. & RAKOV, A.I. (1966) [The state of persons working in electrical fields of outdoor 400- and 500-kV switchyards.] Gig. Tr. Prof. Zabol., 10: 50-52 (in Russian).

BAKER, R.R. (1980) Goal orientation by blindfolded humans after long distance displacement: possible involvement of a magnetic sense. Science, 210: 555-557.

BAKER, R.R., MATHER, J.G., & KENNAUGH, J.H. (1983) Magnetic bones in human sinuses. Nature (Lond.), 310: 78-80.

BARKER, A.T., DIXON, R.A., SHARRAD, W.J.W., & SUTCLIFFE, U.L. (1984) Pulsed magnetic field therapy for tibial non-union. Interim results of a double-blind trial. Lancet, 1: 994-996.

BARLOW, H.B., KOHN H.I., & WALSH, E.G. (1947a) The effect of dark adaption and of light upon the electric threshold of the human eye. Am. J. Physiol., 148: 376-381.

BARLOW, H.B., KOHN, H.I., & WALSH, E.G. (1947b) Visual sensations aroused by magnetic fields. Am. J. Physiol., 148: 372-375.

BARNOTHY, J.M. (1964a) Development of young mice, In: Barnothy, M.F., ed. Biological effects of magnetic fields, New York, London, Plenum Press, Vol. I, pp. 93-99.

BARNOTHY, J.M. (1964b) Rejection of transplanted tumors in mice. In: Barnothy, M.F., ed. Biological effects of magnetic fields, New York, London, Plenum Press, Vol. 1, pp. 100-108.

BARNOTHY, J.M., BARNOTHY, M.F., & BOSZORMENYI-NAGY, I. (1956) Influence of a magnetic field upon the leukocytes of the mouse. Nature (Lond.), 177: 577-578.

BARNOTHY, M.F., ed. (1964) Biological effects of magnetic fields, New York, London, Plenum Press, Vol. 1, p. 324.

BARNOTHY, M.F., ed. (1969) Biological effects of magnetic fields, New York, London, Plenum Press, Vol. 2, p. 314.

BARNOTHY, M.F. & SUMEGI, I. (1969a) Effects of the magnetic field on internal organs and the endocrine system of mice. In: Barnothy, M.F., ed. Biological effects of magnetic fields, New York, New York, London, Plenum Press, Vol. 2, pp. 103-126.

BARNOTHY, M.F. & SUMEGI, I. (1969b) Abnormalities in organs of mice induced by a magnetic field. Nature (Lond.), 221: 270-271.

BARREGARD, L., JARVHOLM, B., & UNGETHUM, E. (1985) Cancer among workers exposed to strong static magnetic fields. Lancet, 2(8460): 892.

BASSETT, C.A.L. (1982) Pulsing electromagnetic fields: a new method to modify cell behaviour of calcified and non-calcified tissues. Calcif. Tissue Int., 34: 1-8.

BASSETT, C.A.L., PAWLUK, R.J., & PILLA, A.A. (1974) Augmentation of bone repair by inductively coupled electromagnetic fields. Science, 184: 575-577.

BASSETT, C.A.L., PILLA, A.A., & PAWLUK, R.J. (1977) A non-operative salvage of surgically resistant pseudoarthroses and non-unions by pulsing electromagetic fields. Clin. Orthop. relat. Res., 124: 128-143.

BASSETT, C.A.L., MITCHELL, S.N., & GASTON, S.R. (1982) Pulsing electromagnetic field treatment in ununited fractures and failed arthrodeses. J. Am. Med. Assoc., 247: 623-628.

BATKIN, S. & TABRAH, F.L. (1977) Effects of alternating magnetic field (12 Gauss) on transplanted neuroblastoma. Res. Commun. chem. Pathol. Pharmacol., 16: 351-362.

BATTOCLETTI, J.H., SALLES-CUNHA, S., HALBACH, R.E., NELSON, J., SANCES, A., Jr, & ANTONICH, F.J. (1981) Exposure of rhesus monkeys to 20 000G steady magnetic field: effect on blood parameters. Med. Phys., 8: 115-118.

BAUM, J.W., SCHAIRER, I.A., & LINDAHL, K.L. (1979) Tests in the plant Tradescantia for mutagenic effects of strong fields. In: Tenforde, T.S., ed. Magnetic field effects on biological systems, New York, London, Plenum Press, pp. 22-24.

BAWIN, S.M. & ADEY, W.R. (1976) Sensitivity of calcium binding in cerebral tissue to weak environmental electric fields oscillating at low frequency. Proc. Natl. Acad. Sci. (USA), 73: 1999-2004.

BAWIN, S.M., KACZMAREK, L.K., & ADEY, W.R. (1975) Effects of modulated VHF fields on the central nervous system. Ann. N.Y. Acad. Sci., 247: 74-81.

BAWIN, S.M., SHEPPARD, A., & ADEY, W.R. (1978) Possible mechanisms of weak electromagnetic field coupling in brain tissue. Bioelectrochem. Bioenergy, 5: 67-76.

BAWIN, S.M., SHEPPARD, A.R., MAHONEY, M.D., & ADEY, W.R. (1984) Influences of sinusoidal electric fields on excitability in rat hippocampal slice. Brain Res., 323: 227-237.

BAWIN, S.M., SHEPPARD, A.R., MAHONEY, M.D., ABU-ASSAL, M., & ADEY, W.R. (1986) Comparison between the effects of extracellular direct and sinusoidal currents on excitability in hippocampal slices. Brain Res., 362: 350-354.

BECKER, J.F., GEACINTOV, N.E., VAN NOSTRAND, F., & VAN METTER, R. (1973) Orientation of chlorophyll in vivo. Studies with magnetic field oriented Chlorella. Biochem. biophys. Res. Commun., 51: 597.

BECKER, J.F., GEACINTOV, N.E., & SWENBERG, C.E. (1978a) Photovoltages in suspensions of magnetically oriented chloroplasts. Biochim. Biophys. Acta, 503: 545-554.

BECKER, J.F., TRENTACOSTI, F., & GEACINTOV, N.E. (1978b) A linear dichroism study of the orientation of aromatic protein residues in magnetically oriented bovine rod outer segments. Photochem. Photobiol., 27: 51-54.

BECKER, VON, G. (1979) The influence of local geomagnetic and other physical conditions on various insects in an experimental area. Ztg angew. Zool., 66: 391-416.

BEISCHER, D.E. (1962) Human tolerance to magnetic fields. Astronautics, 42: 24-25.

BEISCHER, D.E. (1969) Vectorcardiogram and aortic blood flow of squirrel monkeys (Saimiri sciureus) in a strong superconductive electromagnet. In: Barnothy, M., ed. Biological effects of magnetic fields, New York, London, Plenum Press, Vol. 2, pp. 241-259.

BEISCHER, D.E. & BREHL, R.J. (1975) Search for effects of 45-Hz magnetic fields on liver triglycerides in mice, Pensacola, Florida, Naval Aerospace Medical Research Laboratory (Report No. NAMRL-1197).

BEISCHER, D.E. & KNEPTON, J.C., Jr (1964) Influence of strong magnetic fields on the electrocardiogram of squirrel monkeys (Saimiri sciureus). Aerosp. Med., 35: 939-944.

BEISCHER, D.E. & KNEPTON, J.C., Jr (1966) The electroencephalogram of the squirrel monkey (Saimiri sciureus) in a very high magnetic field, Pensacola, Florida, Naval Aerospace Medical Research Laboratory (NASA Order No. R-39).

BEISCHER, D.E., GRISSETT, J.D., & MITCHELL, R.R. (1973) Exposure of man to magnetic fields alternating at extremely low frequency, Pensacola, Florida, Naval Aerospace Medical Research Laboratory (Report No. NAMRL-1180).

BELLOSSI, A. (1983) No effect of a static uniform magnetic field on mouse trypanosomiasis. Radiat. environ. Biophys., 22: 311-313

BELLOSSI, A., SUTTER-DUB, M.TH., & SUTTER, B.CH.J. (1984) Effects of constant magnetic fields in rats and mice: a study of weight. Aviat. Space environ. Med., 55: 725-730.

BELTRAME, F., CHIABRERA, A., GRATTAROLA, M., GUERINI, P., PARODI, G., PONTA, D., VERNAZZA, G., & VIVANI, R. (1980) Electromagnetic control of cell function. Alta Freq., 49: 101-114.

BENKOVA, N.P. (1975) The magnetic field of the earth and its variation. In: Physical, mathematical and biological problems of effects of electromagnetic fields and ionization of air, Moscow, Nauka, pp. 13-24.

BERNHARDT, J. (1979) The direct influence of electromagnetic fields on nerve and muscle cells in man within the frequency range of 1 Hz and 30 MHz. Radiat. environ. Biophys., 16: 309-329.

BERNHARDT, J.H. (1985) Evaluation of human exposures to low frequency fields In: The impact of proposed radiofrequency radiation standards on military operations, Neuilly-sur-Seine, North Atlantic Treaty Organization, pp. 8.1-8.18, (AGARD Lecture Series No. 138).

BERNHARDT, J.H. (1986) Assessment of experimentally observed bioeffects in view of their clinical relevance and the exposure at work places. In: Bernhardt, J., ed. Biological effects of static and extremely low frequency magnetic fields, Munich, MMV Medizin Verlag, pp.157-168.

BERNHARDT, J.H. & KOSSEL, F. (1984) [Health risks from NMR-tomography and in vivo NMR spectroscopy.] Fortschr. Geb. Röntgenstrahlen Nukl. Med., 141: 251-258 (in German).

BERNHARDT, J.H. & KOSSEL, F. (1985) Recommendations for the safe use of NMR equipment. Clin. phys. physiol. Meas., 6(1): 65-74.

BERNHARDT J.H., HAUBRICH, H.J., NEWI, G., KRAUSE, N., & SCHNEIDER, K.H. (1986) Limits for electric and magnetic fields in DIN VDE standards: considerations for the range 0 to 10 kHz. In: Proceedings of the 1986 International Conference on Large High Voltage Systems, Paris, CIGRE.

BIGLIANI, L.U., ROSENWASSER, M.P., CAULO, N., SCHINK, M.M., & BASSETT, C.A.L. (1983) The use of pulsing electromagnetic fields to achieve arthrodesis of the knee following failed total knee arthroplasty. A preliminary report. J. Bone jt Surg., 65: 480.

BISTOLFI, F., ed. (1983) [Magnetic fields in medicine,] Torino, Minerva Medica, p. 399 (in Italian).

BLACKMAN, C.F., ELDER, J.A., WEIL, C.M., BENANE, S.G., EICHINGER, D.C., & HOUSE, D.E., (1979) Induction of calcium-ion efflux from brain tissue by radio-frequency radiation: effects of modulation frequency and field strength. Radio Sci., 14: 93-98.

BLACKMAN, C.F., BENANE, S.G., ELDER, J.A., LAMPE, J.A., & FAULK, J.M. (1980) Induction of calcium-ion efflux from brain tissue by radiofrequency radiation: effect of sample number and modulation. Frequency of the power density window. Bioelectromagnetics, 1: 35-43.

BLACKMAN, C.F., BENANE, S.G., HOUSE, D.E., JOINES, W.T., & SPEIGEL, R.J. (1982) Further developments of ELF-induced changes in calcium-ion efflux from brain tissues. In: Proceedings of the 4th Annual Meeting of the Bioelectromagnetics Society, Los Angeles, June-July, 1982, Gaithersburg, Maryland, Bioelectromagnetics Society (Abstract F-4).

BLACKMAN, C.F., BENANE, S.G., HOUSE, D.E., & JOINES, W.T. (1985a) Effects of ELF (1-120 Hz) and modulated (50 Hz) fields on the efflux of calcium ions from brain tissue in vitro. Bioelectromagnetics, 6: 1-11.

BLACKMAN, C.F., BENANE, S.G., RABINOWITZ, J.R., HOUSE, D.E., & JOINES, W.T. (1985b) A role for the magnetic field in the radiation-induced efflux of calcium ions from brain tissue in vitro. Bioelectromagnetics, 6: 327-337.

BLAKEMORE, R. (1975) Magnetotactic bacteria. Science, 190: 377-379.

BLAKEMORE, R.P., FRANKEL, R.B., & KALMIJN, A.J. (1980) South-seeking magnetotactic bacteria in the southern hemisphere. Nature (Lond.), 286: 384-385.

BLANKENSHIP, R.E.M., SCHAAFSMA, T.J., & PARSON, W.W. (1977) Magnetic field effects on radical pair intermediates in bacterial photosynthesis. Biochem. Biophys. Acta, 461: 297-305.

BLISS, V.L. & HEPNER, F.H. (1976) Circadian activity rhythm influenced by near zero magnetic field. Nature (Lond.), 261: 411-412.

BOGOLYUBOV, V.M. (1981) [Status and perspective of investigations on biological and clinical effects of magnetic fields.] J. Vopr. kurortol. phisioter. lech. phizkul'tury, 4: 1-5 (in Russian).

BOOKMAN, M.A. (1977) Sensitivity of the homing pigeon to an earth-strength magnetic field. Nature (Lond.), 267: 340-342.

BORE, P.J. (1985) Safety of NMR. Lancet, 8437: 1107-1108.

BOROSKE, E. & HELFRICH, W. (1978) Magnetic aresotropy of egg lecithin membranes. Biophys. J., 24: 863-868.

BRETON, J. (1974) The state of chlorophyll and carotenoid in vivo. II-A linear dichroism study of pigment orientation in photosynthetic bacteria. Biochem. biophys. Res. Commun., 59: 1011-1017.

BREWER, H.B. (1979) Some preliminary studies of the effects of a static magnetic field on the life cycle of the Lebistes reticulatus (guppy). Biophys. J., 28: 305-314.

BROWN, F.A., Jr & SCOW, K.M. (1978) Magnetic induction of a circadian cycle in hamsters. J. Interdiscipl. Cycle Res., 9: 137-145.

BUBENIK, G.A., PURTRILL, R.A., BROWN, G.M., & GROTA, L.J. (1978) Melatonin in the retina and the Harderian gland: ontogeny, diurnal variations and melatonin treatment. Exp. Eye Res., 27: 323-333.

BUCKING, J., HERBST, M., & PIONTEK, P. (1974) The influence of a strong magnetic field on muscular contraction. Radiat. environ. Biophys., 11: 79-85.

BUDD, R.A. & CZERSKI, P. (1985) Modulation of mammalian immunity by electromagnetic radiation (EMR). J. Microwave Power, 20: 217-231.

BUDINGER, T.F. (1979) Thresholds for physiological effects due to RF and magnetic fields used in NMR imaging. IEEE Trans. Nucl. Sci., 26(2): 2821-2825.

BUDINGER, T.F. (1981) Nuclear magnetic resonance (NMR) in in vivo studies: known thresholds for health effects. J. Comput. assisted Tomogr., 5(6): 800-811.

BUDINGER, T.F. & CULLANDER, C. (1983) Biophysical phenomena and health hazards of in vivo magnetic resonance. In:

Margulis, A.R., Higgins, C.B., Kaufman, L., & Crooks, L.E., ed. Clinical magnetic resonance imaging, San Francisco, University of California Press, pp. 303-320.

BUDINGER, T.F. & LAUTERBUR, P.C. (1984) Nuclear magnetic resonance technology for medical studies. Science, 226: 288-298.

BUDINGER, T.F., CULLANDER, C., & BORDOW, R. (1984a) Switched magnetic field thresholds for the induction of magnetophosphenes (Presented at the 3rd Annual Meeting of the Society for Magnetic Resonance in Medicine, New York, 4-6 August).

BUDINGER, T.F., BRISTOL, K.S., YEN, C.K., & WONG, P. (1984b) Biological effects of static magnetic fields (Presented at the 3rd Annual Meeting of the Society for Magnetic Resonance in Medicine, New York, 4-6 August).

BUREAU OF RADIOLOGICAL HEALTH (1981) An evaluation of radiation emission from video display terminals, Rockville, Mayrland, Bureau of Radiological Health (HHA-FDA 81-8153).

BUTLER, B.C. & DEAN, W.W. (1964) The inhibitory effect of a magnetostatic field upon the tissue growth of KB cells. Am. J. Med. Electron., 3: 123-125.

BUTROUS, G.S., MALE, J.C., WEBER, R.S., BARTON, D.G., MELDRUM, S.J., BONNELL, J.A., & CAMM, A.J. (1983) The effect of power frequency high intensity electric fields on implanted cardiac pacemakers. Pacing clin. Electrophysiol., 6: 1282-1292.

BUYAVIKH, A.G. (1984) [Effect of low-frequency magnetic field upon lipoprotein blood spectrum of a patient with hypertension.] Vrach. Delo, 9: 75-77 (in Russian).

CAIN, C.S., LUBEN, R.A., & ADEY, W.R. (1984) Pulsed electromagnetic field effects on PTH stimulated cAMP accumulation and bone resorption in mouse calvariae (Presented at the 23rd Annual Hanford Life Sciences Symposium on Interaction of Biological Systems with Static and ELF Electric and Magnetic Fields, Richland, Washington, 2-4 October).

CAIN, C.S., DONATO, N.J., BYUS, C.V., ADEY, W.R., & LUBEN, R.A. (1985) Pulsed electromagnetic field influences PTH-stimulated cAMP accumulation and ornithine decarboxylase activity in primary bone marrow cells. In: Proceedings of the US Department of Energy Contractors Review Meeting, Arlington, Virginia, 6-11 November, 1985.

CALDWELL, W.E. & RUSSO, F. (1968) An exploratory study of the effects of an AC magnetic field upon the behaviour of the Italian Honey bee (Apis mellifica). J. Genet. Psych., 113: 233-252.

CALLE, E. & SAVITZ, D.A. (1985) Leukaemia in occupational groups with presumed exposure to electrical and magnetic fields. New Engl. J. Med., 313: 1476-1477.

CDRH (1982) Guidelines for evaluating electromagnetic risk for trials of clinical NMR systems, Rockville, Maryland, Center for Devices and Radiological Health, US Department of Health and Human Services, US Food and Drug Administration (February 25 and November 29).

CHAGNEUX, R. & CHALAZONITIS, N. (1972) Evaluation de l'anisotropie magnétique des cellules multimembranaires dans un champ magnétique constant (segments externes des bâtonnets de la rétine de grenouille). C. R. Acad. Sci. (Paris) Ser. D, 274: 317-320.

CHAGNEUX, R., CHAGNEUX, H., & CHALAZONITIS, N. (1977) Decrease in magnetic anisotropy of external segments of the retinal rods after a total photolysis. Biophys. J., 18: 125-127.

CHALAZONITIS, R., CHAGNEUX, R., & ARVANITAKI, A. (1970) Rotation des segments externes des photorécepteurs dans le champ magnétique constant, C. R. Acad. Sci. Paris Ser. D, 271: 130-133.

CHAN, C.Y. & NICHOLSON, C. (1986) Modulation by applied electric fields of purkinje and stellate cell activity in the isolated turtle cerebellum. J. Physiol., 371: 89-114.

CHANDRA, S. & STEFANI, S. (1979) Effect of constant and alternating magnetic fields on tumour cells in vitro and in vivo. In: Phillips, R.D., Gillis, M.F., Kaune, W., & Mahlum, D.D., ed. Biological effects of extremely low frequency electromagnetic fields, Springfield, Virginia, US Department of Energy, National Technical Information Service, pp. 436-446 (NTIS Report No. CONF-781016).

CHIABRERA, A., GRATTOVOLA, M., VERNAZZA, G., & VIVIANI, R. (1978) Bioelectrochemical system models, electromagnetic interactions and noise. Bioelectrochem. Bioenergy, 5: 97-115.

CHIABRERA, A., HINSENKAMP, M., PILLA, A.A., RYABY, J., PONTA, D., BELMONT, A., BELTRAME, F., GRATTAROLO, M., & NICOLINI, C.

(1979) Cytofluorometry of electromagnetically controlled cell dedifferentiation. J. Histochem. Cytochem., 27: 375-381

CHIABRERA, A., GRATTAROLA, M., & VIVIANI, R. (1984) Interaction between electromagnetic fields and cells: microelectrophoretic effect on ligands and surface receptors. Bioelectromagnetics, 5: 173-191.

CLARKE, R.L. & JUSTESEN, D.R. (1979) Behavioural sensitivity of a domestic bird to 60 Hz AC and to DC magnetic fields. Radio Sci., 14(6S): 209-216.

COLEMAN, M., BELL, J., & SKEET, R. (1983) Leukaemia incidence in electrical workers. Lancet, 1: 982-983.

COLEMAN, M., BELL, C.M.J., TAYLOR, H.L., & THORNTON-JONES, H. (1985) Leukaemia and electromagnetic fields: a case control study (Paper presented at the International Conference on Electric and Magnetic Fields in Medicine and Biology, London, October, 1985).

CONTI, P., GIGANTE, G.E., CIFONE, M.G., ALESSE, E., IANNI, G., REALE, M., & ANGELETTI, P.U. (1983) Reduced mitogenic stimulation of human lymphocytes by extremely low frequency electromagnetic fields. FEBS Lett., 162: 156-160.

CONTI, R. (1985) Instrumentation for measurements of power frequency electromagnetic fields. In: Grandolfo, M., Michaelson, S.M., & Rindi, A., ed. Biological effects and Dosimetry of static and ELF electromagnetic fields, New York, London, Plenum Press, pp. 187-210.

COOK, E.S. & SMITH, M.J. (1964) Increase of trypsin activity, In: Barnothy, M.F., ed. Biological effects of magnetic fields, New York, London, Plenum Press, Vol. 1, pp. 246-254.

COOK, E.S., FARDON, J.C., & NUTINI, L.G. (1969) Effects of magnetic fields on cellular respiration. In: Barnothy, M.F., ed. Biological effects of magnetic fields, New York, London, Plenum Press, Vol. 2, pp. 67-78.

COPE, F.W. (1971) Evidence from activation energies from superconductive tunnelling in biological systems at physiological temperatures. Physiol. Chem. Phys., 3: 403-410.

COPE, F.W. (1973) Biological sensitivity to weak magnetic fields due to biological superconductive Josephson junctions, Physiol. Chem. Phys., 5: 173-176.

COPE, F.W. (1974) Enhancement by high electric fields of superconduction in organic and biological solids at room temperature and a role in nerve conduction. Physiol. Chem. Phys., 6: 405-410.

COPE, F.W. (1978) Discontinuous magnetic field effects (Barkhausen noise) in nucleic acids as evidence for room temperature organic superconduction. Physiol. Chem. Phys., 10: 233-246.

COPE, F.W. (1981) On the relativity and uncertainty of electromagnetic energy measurement at a superconductive boundary. Application to perception of weak magnetic fields by living systems. Physiol. Chem. Phys., 13: 231.

CREIM, J.A., LOVELY, R.H., KAUNE, W.T., & PHILLIPS, R.D. (1984) Exposure to 30-Gauss magnetic fields does not cause avoidance behaviour in rats (Presented at the 23rd Annual Hanford Life Sciences Symposium on Interaction of Biological Systems with Static and ELF Electric and Magnetic Fields, Richland, Washington, 2-4 October).

CREIM, J.A., LOVELY, R.H., KAUNE, W.T., MILLER, M.C., & ANDERSON, L.E. (1985) 60 Hz magnetic fields: do rats avoid exposure? (Presented at the 7th Annual Meeting of the Bioelectromagnetics Society, San Francisco, California, 16-20 June).

CROOKS, L.E. & KAUFMAN, L. (1983) Basic physical principles and imaging techniques. In: Margulis, A.R., Higgins, C.B., Kaufman, L., & Crooks, L.E., ed. Clinical magnetic resonance imaging, San Francisco, University of California Press, pp. 13-30.

CZERSKI, P. (1986) Experimental observations on bio-effects of static and time varying magnetic fields. In: Bernhardt, J.H., ed. Biological effects of static and extremely low frequency magnetic fields, Munich, MMV Medizin Verlag, pp. 75-85.

DAVIS, H.P., MIZUMORI, J.Y., ALLEN, H., ROSENZWEIG, M.R., BENNETT, E.L., & TENFORDE, T.S. (1984) Behavioural studies with mice exposed to DC and 60 Hz magnetic fields. Bioelectromagnetics, 5: 147-164.

D'ARSONVAL, M.A. (1896) Dispositifs pour la mesure des courants alternatifs à toutes fréquences. C. R. Soc. Biol. (Paris), 3: 450-451.

DELGADO, J.M.R., MONTEAGUDO, J.L., GARCIA, M.G., & LEAL, J. (1981) Teratogenic effects of weak magnetic fields. IRCS Med. Sci., 9: 392.

DELGADO, J.M.R., MONTEAGUDO, J.L., GARCIA, M.G., & LEAL, J. (1982) Embryological changes induced by weak, extremely low frequency electromagnetic fields. J. Anat., 134: 533-551.

DELGADO, J.M.R., MONTEAGUDO, J.L., & RAMIREZ, (1983) Non-invasive magnetic stimulation of the monkey cerebellum (Presented at the 5th Annual Meeting of the Bioelectromagnetics Society, Boulder, Colorado, 12-16 June).

DELORGE, J. (1972) Operant behaviour of rhesus monkeys in the presence of low-frequency low-intensity magnetic and electric fields: experiment 1, Pensacola, Florida, Naval Aerospace Medical Research Laboratory (Report No. NAMRL-1155) (NTIS No. AD754058).

DELORGE, J. (1973a) Operant behaviour of rhesus monkeys in the presence of low-frequency low-intensity magnetic and electric fields: experiment 2, Pensacola, Florida, Naval Aerospace Medical Research Laboratory (Report No. NAMRL-1179) (NTIS No. AD764532).

DELORGE, J. (1973b) Operant behaviour of rhesus monkeys in the presence of low-frequency low-intensity magnetic and electric fields: experiment 3, Pensacola, Florida, Naval Aerospace Medical Research Laboratory (Report No. NAMRL-1196 (NTIS No. AD774106).

DELORGE, J. (1974) A psychobiological study of rhesus monkeys exposed to extremely low-frequency low-intensity magnetic fields, Pensacola, Florida, Naval Aerospace Medical Research Laboratory (Report No. NAMRL-1203) (NTIS No. AD/A000078).

DELORGE, J. (1979) Effects of magnetic fields on behaviour in nonhuman primates. In: Tenforde, T.S., ed. Magnetic field effects on biological systems, New York, London, Plenum Press, pp. 37-38.

DELORGE, J. (1985) Behavioural studies on monkeys in electric and magnetic fields at ELF frequencies. In: Grandolfo, M., Michaelson, S.M., & Rindi, A., ed. Biological effects and dosimetry of static and ELF electromagnetic fields, New York, London, Plenum Press, pp. 433-439.

DEMETSKY, A.M. & ALEKSEEV, A.G. (1981) [Man-made magnetic fields in medical practice (experimental studies),] Minsk, Izd-vo Belarus, p. 94 (in Russian).

DEMETSKY, A.M., SURGANOVA, S.F., POPOVA, L.I., & GAVILOVICH, P.F. (1979) Magnetic field effects on microcirculation. Zdravookhr Belorus, 12: 3-5.

DENISOV, V.N., GAVRILOV, V.A., & INNONIKOV, I.N. (1979) Human reactions to the influence of electric and magnetic fields at industrial frequency. In: Living systems in electromagnetic fields, Tomsk, University of Tomsk, pp. 54-64.

D'SOUZA, L., RENO, V.R., NUTINI, L.G., & COOK, E.S. (1969) The effect of a magnetic field on DNA synthesis by ascites Sarcoma 37 cells. In: Barnothy, M.F., ed. Biological effects of magnetic fields, New York, London, Plenum Press, Vol. 2, pp. 53-59.

DIEBOLT, J.R. (1978) The influence of electrostatic and magnetic fields on mutation in Drosophila melanogaster spermatozoa. Mutat. Res., 57: 169-174.

DIXEY, R. & REIN, G. (1982) ^{3}H-noradrenaline release potentiated in a clonal nerve cell line by low-intensity pulsed magnetic fields. Nature (Lond.), 296: 253-256.

DRUZ, V.A. & MADIYEVSKII, YU.M. (1966) [Effects of constant magnetic and low-frequency electromagnetic fields on the hydration capacity of surviving tissues.] Biofizika, 11: 724-731 (in Russian with English abstract).

DUNLAP, K. (1911) Visual sensations from the alternating magnetic field. Science, 33: 68-71.

EDELMAN, A., TEULEN, J., & PUCHALSKA, I.B. (1979) Influence of magnetic fields on frog sciatic nerve. Biochem. biophys. Res. Commun., 91: 118-122.

EISELEIN, B.S., BOUTELL, H.M., & BIGGS, M.W. (1961) Biological effects of magnetic fields - negative results, Aerosp. Med., 32: 383-386.

EMLEN, S.T., WILTSCHKO, W., DEMONG, N.J., WILTSCHKO, R., & BERGMAN, S. (1976) Magnetic direction finding: evidence for its use in migratory indigo buntings. Science, 193: 505.

ERLEBACHER, J.A., CAHILL, P.T., PANNIZZO, F., & KNOWLES, R.J.R. (1986) Effects of magnetic resonance imaging on DDD pacemakers. Am. J. Cardiol., 57: 437-440.

FAM, W.Z. (1981) Biological effects of 60-Hz magnetic field on mice. IEEE Trans. Mag., 17: 1510-1513.

FEDERAL REPUBLIC OF GERMANY (1984) Hazards of electromagnetic fields. Protection of persons in the frequency range 10 kHz to 3000 GHz, Berlin, Deutsche Norm (DIN VDE No. 0848, July).

FEDERAL REPUBLIC OF GERMANY (1986) Hazards from electromagnetic fields. Protection of persons in the frequency range from 0 Hz to 3000 GHz, Berlin, Deutsche Norm (New DIN-VDE draft, June).

FHO (1984) [Recommendations on avoiding health risks caused by magnetic and radiofrequency electromagnetic fields during nuclear magnetic resonance tomography and in vivo nuclear magnetic resonance spectroscopy.] Bundesgesundheitsblatt, 27(3): 92-96 (in German).

FOSTER, M.A. (1984) Magnetic resonance in medicine and biology, Oxford, Pergamon Press.

FRANKEL, R.B. (in press) Biological effects of static magnetic fields. In: Polk, C. & Postow, E., ed. Handbook of biological effects of electromagnetic fields, Boca Raton, Florida, CRC Press.

FRANKEL, R.B., BLAKEMORE, R.P., & WOLFE, R.S. (1979) Magnetite in fresh-water magnetotactic bacteria. Science, 203: 1355-1356.

FRANKEL, R.B., BLAKEMORE, R.P., TORRES DE ARAUJO, F.F., & ESQUIVAL, D.M.S. (1981) Magnetotactic bacteria at the geomagnetic equator. Science, 212: 1269-1270.

FRAZIER, M.E., ANDREWS, T.K., & THOMPSON, B.B. (1979) In vitro evaluation of biomagnetic effects. In: Phillips, R.D., Gillis, M.F., Kaune, W.T., & Mahlum, D.D., ed. Biological effects of extremely low frequency electromagnetic fields, Springfield, Virginia, US Department of Energy, National Technical Information Service, pp. 417-465 (NTIS Rep. No. CONF-781016).

FRIEDMAN, H. & CAREY, R.J. (1972) Biomagnetic stressor effects in primates. Physiol. Behav., 9: 171-173.

FRIEDMAN, H., BECKER, R.O., & BACHMAN, C.H. (1967) Effect of magnetic fields on reaction time performance. Nature (Lond.), 213: 949-950.

FROHLICH, H. (1968) Long-range coherence and energy storage in biological systems. Int. J. Quantum Chem., 2: 641-649.

FROHLICH, H. (1977) Possibilities of long- and short-range electric interactions of biological systems. Neurosci. Res. Program Bull., 15: 67-72.

FUJITA, T.Y. & TENFORDE, T.S. (1982) Portable magnetic field dosimeter with data acquisition capabilities. Rev. sci. Instrum., 53: 326-331.

FULTON, J.P., COBB, S., PREBLE, L., LEONE, L., & FORMAN, E. (1980) Electrical wiring configurations and childhood leukaemia in Rhode Island. Am. J. Epidemiol., 113(3): 292-296.

GAFFEY, C.T. & TENFORDE, T.S. (1979) Changes in the electrocardiograms of rats and dogs exposed to DC magnetic fields, Berkeley, University of California, Lawrence Berkeley Laboratory (Report No. 9085).

GAFFEY, C.T. & TENFORDE, T.S. (1981) Alterations in the rat electrocardiogram induced by stationary magnetic fields. Bioelectromagnetics, 1: 357-370.

GAFFEY, C.T. & TENFORDE, T.S. (1983) Bioelectric properties of frog sciatic nerves during exposure to stationary magnetic fields. Radiat. environ. Biophys., 22: 61-73.

GAFFEY, C.T. & TENFORDE, T.S. (1984) Electroretinograms of cats and monkeys exposed to large stationary magnetic fields (Presented at the 6th Annual Meeting of the Bioelectromagnetics Society, Atlanta, Georgia, 15-19 July).

GAFFEY, C.T., TENFORDE, T.S., & DEAN, E.E. (1980) Alterations in the electrocardiograms of baboons exposed to DC magnetic fields. Bioelectromagnetics, 1: 209.

GALAKTIONOVA, G.V. & STRZHIZHOVSKY, A.G. (1973) [Effect of permanent magnetic fields up to 4500 Oe on the mitotic activity of corneal epithelial cells in mice.] Kosm. biol. aviakosm. Med., 7: 49-51 (in Russian).

GALAKTIONOVA, G.V., MASTRYUKOVA, M., & STRZHIZHOVSKY, A.D. (1985) [Sensitivity of mammalian tissues to prolonged effects of constant magnetic fields of high strength.] Kosm. biol. aviakosm. Med., 19: 78-81 (in Russian).

GAUGER, J.R. (1984) Household appliance magnetic field survey, Arlington, Virginia, Naval Electronic Systems Command, (IIT Research Institute Report EO 6549-3).

GEACINTOV, N.E., VAN NOSTRAND, F., POPE, M., & TINKEL, J.B. (1971) Magnetic field effect on the chlorophyll fluorescence in Chlorella. Biochim. Biophys. Acta, 226: 486-491.

GEACINTOV, N.E., VAN NOSTRAND, F., BECKER, J.F., & TINKEL, J.B. (1972) Magnetic field induced orientation of photosynthetic systems. Biochim. Biophys. Acta, 267: 65-79.

GERENCER, V.F., BARNOTHY, M.F., & BARNOTHY, J.M. (1962) Inhibition of bacterial growth by magnetic fields. Nature (Lond.), 196: 539-541.

GERMAIN, C. (1963) Bibliographical review of the methods of measuring magnetic fields. Nucl. Instrum. Method, 21: 17-46.

GIBSON, R.F. & MORONEY, W.F. (1974) The effect of extremely low frequency magnetic fields on human performance: a preliminary study, Pensacola, Florida, Naval Aerospace Medical Research Laboratory (Report No. NAMRL-1195).

GILMAN, P.A., AMES, R.G., & MCCAWLEY, M.A. (1985) Leukaemia risk among US white male coal miners. J. occup. Med., 27: 669-671.

GOLDBERG, V.E. & MEL'NIK-GUYKAZYAN, E.V. (1980) [The effect of 50 Hz, 200 oersted variable magnetic field upon hemopoiesis in mice at 24-hour exposure.] In: [Problems of theoretical and clinical medicine,] Tomsk, USSR Academy of Medical Sciences, pp. 12-15 (in Russian).

GOODMAN, E.M., GREENEBAUM, B., & MARRON, M.T. (1979) Bioeffects of extremely low-frequency electromagnetic fields: variation with intensity, waveform, and individual or combined electric and magnetic fields. Radiat. Res., 78: 485-501.

GOODMAN, R. & HENDERSON, A.S. (1986) Sine waves enhance cellular transcription. Bioelectromagnetics, 7: 23-30.

GOODMAN, R., BASSETT, C.A.L., & HENDERSON, A.S. (1983) Pulsing electromagnetic fields induce cellular transcription. Science, 220: 1283-1285.

GORCHONSKAYA, E. (1984) [Effect of static magnetic field upon hemorrhage.] Kosm. biol. aviakosm. Med., 18(6): 87-89 (in Russian).

GORCZYNSKA, E., GALKA, G., KROLIKOWSKA, R., & WEGRZYNOWICZ, R. (1982) Effect of magnetic field on activity of cytochrome oxidase not moved or moved relative to magnetic field lines. Physiol. Chem. Phys., 14: 201-207.

GOULD, J.L. & ABLE, K.P. (1981) Human homing: an elusive phenomenon. Science, 214: 1061-1063.

GOULD, J.L., KIRSCHVINK, J.L., & DEFFEYES, K.S. (1978) Bees have magnetic resonance. Science, 201: 1026-1028.

GRAHAM, C., COHEN, H.D., COOK, M.R., PHELPS, J., GERKOVICH, M., & FOTOPOULES, S.S. (1984) A double-blind evaluation of 60-Hz field effect on human performance, physiology, and subjective state (Presented at the 23rd Annual Hanford Life Sciences Symposium on Interaction of Biological Systems with Static and ELF Electric and Magnetic Fields, Richland, Washington, 2-4 October).

GRAHAM, J., TESCH, C., & WENDHAUSEN, H. (1986) [Biological reaction of men and animals in static magnetic fields.] In: [Current problems concerning the physics, technology and biology of NMR in medicine,] Stuttgart, Urban and Schwarzenberg (in German).

GRANDOLFO, M. & VECCHIA, P. (1985a) Physical description of exposure to static and ELF electromagnetic fields. In: Grandolfo, M., Michaelson, S.M., & Rindi, A., ed. Biological effects and dosimetry of static and ELF electromagnetic fields, New York, London, Plenum Press, pp. 31-48.

GRANDOLFO, M. & VECCHIA, P. (1985b) Natural and man-made environmental exposures to static and ELF electromagnetic fields. In: Grandolfo, M., Michaelson, S.M., & Rindi, A., ed. Biological effects and dosimetry of static and ELF electromagnetic fields, New York, London, Plenum Press, pp. 49-70.

GREENBERG, B., BINDOKAS, V.P., & GAUGER, J.R. (1981a) Biological effects of a 765-kV transmission line: exposures and thresholds in Honey bee colonies. Bioelectromagnetics, 2: 315-328.

GREENBERG, B., BINDOKAS, V.P., FRAZIER, M.J., & GAUGER, J.R. (1981b) Response of Honey bees Apis mellifera L. to high-voltage transmission lines. Environ. Entomol., 10: 600-610.

GREENEBAUM, B., GOODMAN, E.M., & MARRON, M.T. (1979) Extremely low-frequency fields and the slime mold Physarum polycephalum: evidence of depressed cellular function and of internuclear interaction. Radio Sci., 14(6S): 103-107.

GREENEBAUM, B., GOODMAN, E.M., & MARRON, M.T. (1982) Magnetic field effects on mitotic cycle length in Physarum. Eur. J. cell Biol., 27: 156-160.

GREMMEL, H., WENDHAUSEN, H., & WUNSCH, F. (1984) [Biological effects of static magnetic fields associated with NMR tomography in man,] Kiel, University of Kiel, Radiology Department (in German).

GRIFFIN, J.C. (1985) The effects of ELF electric and magnetic fields on artificial cardiac pacemakers. In: Assessments and viewpoints on the biological and human health effects of extremely low frequency (ELF) electromagnetic fields, Washington DC, American Institute of Biological Sciences, pp. 173-183.

GRISSETT, J.D. (1971) Exposure of squirrel monkeys for long periods of extremely low-frequency magnetic fields: central nervous system effects as measured by reaction time, Pensacola, Florida, Naval Aerospace Medical Research Laboratory (Report No. NAMRL-1146).

GRISSETT, J.D. & DELORGE, J. (1971) Central nervous system effects as measured by reaction time in squirrel monkeys exposed for short periods to extremely low-frequency magnetic fields, Pensacola, Florida, Naval Aerospace Medical Research Laboratory (Report No. NAMRL-1137).

GRODSKY, I.T. (1976) Neuronal membrane: a physical synthesis. Math. Biosci., 28: 191-194.

GRODSKY, I.T. (1977) Molecular and physical bases of tissue interaction with electromagnetic fields. Neurosci. Res. Program Bull., 15: 191-219

GROHMANN, J., TESCH, C., & WENDHAUSEN, H. (1986) [Biological reaction of man and animals in static magnetic fields.] In: [Current problems concerning the physics, technology and biology of NMR in medicine,] Stuttgart, Urban & Schwarzenberg (in German).

GROSS, L. (1964) Lifespan increase of tumor-bearing mice through pretreatment. In: Barnothy, M.F., ed. Biological effects of magnetic fields, New York, London, Plenum Press, Vol. 1, pp. 132-139.

GUNDERSEN, R. & GREENEBAUM, B. (1985) The effects of ELF magnetic fields on synaptic function (Presented at the 7th Annual Meeting of the Bioelectromagnetics Society, San Francisco, California, 16-20 June).

HABERDITZL, W. (1967) Enzyme activity in high magnetic fields. Nature (Lond.), 213: 72-73.

HABERHORN, R. & MICHEL-BEYERLE, M.E. (1979) On the mechanism of magnetic field effects in bacterial photosynthesis. Biophys. J., 26: 489-498.

HAKEMI, H. & LABES, M.M. (1974) New optical method for studying anisotropic diffusion in liquid crystals. J. Chem. Phys., 61: 4020-4025.

HAKEMI, H. & LABES, M.M. (1975) Self-diffusion coefficients of a nematic liquid crystal via an optical method. J. Chem. Phys., 63: 3708-3712.

HALL, E.J., BEDFORD, J.S., & LEASK, M.J.M. (1964) Some negative results in the search of a lethal effect of magnetic fields on biological materials. Nature (Lond.), 203: 1086-1087.

HALPERN, M.H. & GREEN, A.E. (1964) Effects of magnetic fields on growth of HeLa Cells in tissue culture. Nature (Lond.), 202: 717.

HANNEMAN, G.D. (1969) Changes in sodium and potassium content of urine from mice subjected to intense magnetic fields, In: Barnothy, M.F., ed. Biological effects of magnetic fields, New York, London, Plenum Press, Vol. 2, pp. 127-136.

HAUF, R. (1976) Influence of 50 Hz alternating electric and magnetic fields on human beings. Rev. gen. electr., Numéro spécial, July: 31-49.

HAUF, R. (1982) Electric and magnetic fields at power frequencies, with particular reference to 50 and 60 Hz. In: Suess, M.J., ed. Non-ionizing radiation protection, Copenhagen, World Health Organization, Regional Office for Europe, pp. 175-197 (WHO Regional Publications European Series No. 10).

HASSENZAHL, W., MAHAFFY, M., & WEINFROFEN, J. (1978) Evaluation of environmental control technologies for magnetic fields, Springfield, Virginia, US Department of Energy, National Technical Information Service (NTIS Report No. DOE/EV-0029).

HCN (1984) Interim advice on NMR, The Hague, Health Council of the Netherlands (19 January).

HEALTH AND WELFARE CANADA (1985) Recommendations to ensure protection of patients and operational personnel from potential hazards in proton NMR imaging, Ottawa, Environmental Health Directorate, p. 17 (Report 85-EHD-124).

HEALTH AND WELFARE CANADA (1986) Guidelines on exposure to electromagnetic fields from magnetic resonance clinical systems, Ottawa, Environmental Health Directorate (EHD Safety Code Report No. 86-EHD-127).

HINSENKAMP, M.G. (1982) Traitement des pseudarthroses par stimulation électromagnétique. Rev. méd. Bruxelles, 3(1): 19-28 (with English abstract).

HINSENKAMP, M.G. & ROOZE, M.A. (1982) Morphological effect of electromagnetic stimulation on the skeleton of fetal or newborn mice. Acta orthop. Scand., Suppl., 196: 39.

HOFF, A.F. (1981) Magnetic field effects on photosynthetic reactions. Q. Rev. Biophys., 14: 599-665.

HONG, F.T. (1977) Photoelectric and magneto-orientation effects in pigmented biological membranes. J. colloid interface Sci., 58: 471-497.

HONG, F.T. (1980) Magnetic anisotropy of the visual pigment rhodopsin. Biophys. J., 29: 343-346.

HONG, F.T., MAUZERALL, D., & MAURO, A. (1971) Magnetic anisotropy and the orientation of retinal rods in a homogeneous magnetic field. Proc. Natl Acad. Sci. (USA), 68: 1283-1285.

ILINSKY, O.P. & BROWN, R.S. (1985) [Electroreceptors in fish,] Leningrad, Nauka (in Russian).

IRPA (1985) Review of concepts, quantities, units and terminology for non-ionizing radiation protection. Health Phys., 49(6): 1329-1362.

IRWIN, D.D., RUSH, S., EVERING, R., LEPESCHKIN, E., MONTGOMERY, D.B., & WEGGEL, R.J. (1970) Stimulation of cardiac muscle by a time-varying magnetic field. IEEE Trans. Mag., 6: 321-322.

IWASAKI, T., OHARA, H., MATSUMOTO, S., & MATSUDAIRA, H. (1978) Test of biological sensitivity in three different biological systems. J. Radiat. Res., 19: 287-294.

JAFARY-ASL, A.H., SOLANKI, S.N., AARHOLT, E., & SMITH, C.W. (1982) Dielectric measurements on live biological materials under magnetic resonance conditions. J. Biol. Phys., 11: 15-22.

JEFFRIES, J.G.R. (1981) Influence of electric fields on the excitability of granule cells in guinea-pig hippocampal slices. J. Physiol., 319: 143-152.

JENKINS, B.M. & WOODY, J.A. (1978) Cardiac pacemaker responses to power frequency signals. In: Proceedings of the IEEE International Symposium on Electromagnetic Compatibility, Piscataway, New Jersey, Institute of Electrical and Electronic Engineering, Vol. S78, pp. 273-277.

JOLLEY, W.B., HINSHAW, D.B., KNIERIM, K., & HINSHAW, D.B. (1983) Magnetic field effects on calcium efflux and insulin secretion in isolated rabbit islets of Langerhans. Bioelectromagnetics, 4: 103-106.

JONES, D.S. & MACFADDEN, B.J. (1982) Induced magnetization in the monarch butterfly *Danaus plexippus* (Insecta, Lepidoptera). J. exp. Biol., 96: 1-9.

JOSHI, M.Y., KHAN, M.Z., & DAMLE, P.S. (1978) Effect of magnetic field on chick morphogenesis. Differentiation, 10: 39-43.

JUUTILAINEN, J. & SAALI, J. (1986) Development of chick embryos in 1 Hz to 100 kHz magnetic fields. Radiat. environ. Biophys., 25: 135-140.

JUUTILAINEN, J., HARRI, M., SAALI, K., & LAHTINEN, T. (1986) Effects of 100-Hz magnetic fields with various waveforms on the development of chick embryos. Radiat. environ. Biophys., 25: 65-74.

KACZMAREK, L. (1977) Cation binding models for the interaction of membranes with EM fields. Neurosci. Res. Program Bull., 15: 54-60.

KALE, P.G. & BAUM, J.W. (1979) Genetic effects of strong magnetic fields in *Drosophila melanogaster*. I. Homogeneous fields ranging from 13 000 to 37 000 gauss. Environ. Mutagen., 1: 371-374.

KALMIJN, A.J. (1974) The detection of electric fields from inanimate and animate sources other than electric organs. In: Autrum, H., Jung, R., Loewenstein, W.R., MacKay, D.M., & Teiber, H.L., ed. Handbook of sensory physiology, Berlin, Heidelberg, New York, Springer-Verlag, pp. 147-200.

KALMIJN, A.J. (1978) Experimental evidence of geomagnetic orientation in elasmobranch fishes. In: Schmidt-Koenig, K. & Keeton, W.T., ed. Animal migration, navigation, and homing, Berlin, Heidelberg, New York, Springer-Verlag, pp. 347-353.

KALMIJN, A.J. (1981) Biophysics of geomagnetic field detection. IEEE Trans. Mag., 17: 1113-1124.

KALMIJN, A.J. (1982) Electric and magnetic field detection in elasmobranch fishes. Science, 218: 916-918.

KALMIJN, A.J. (1984) Theory of electromagnetic orientation: a further analysis. In: Bolis, L., Keynes, R.D., & Maddrell, S.H.P., ed. Comparative physiology of sensory systems, Cambridge, Cambridge University Press, pp. 525-560.

KANDIL, A. & ELASHMAWY, H. (1981) The anti-inflammatory potential of a magnetic band. J. Drug Res., 12: 127-130.

KARTASHEV, A.G., KALYUZHIN, V.A., & MIGALKIN, I.V. (1978) [Effect of weak magnetic field on rate of glycolytic reaction of Saccharomyces cerevisiae.] Kosm. biol. aviakosm. Med., 12: 76-77 (in Russian).

KAUNE, W.T. (1985) Coupling of living organisms to ELF electric and magnetic fields. In: Biological and human health effects of extremely low frequency electromagnetic fields, Arlington, Virginia, American Institute of Biological Sciences, pp. 25-60.

KAUNE, W.T. & CURLEY, R. (1986) Induction of currents in humans and animals by ELF magnetic fields. In: Proceedings of the 8th Annual Meeting of the Bioelectromagnetics Society, Madison, Wisconsin, 1-5 June, Gaithersburg, Maryland, Bioelectromagnetics Society, p. 17.

KAVALIERS, M., OSSENKOPP, K.P., & HIRST, M. (1984) Magnetic fields abolish the enhanced nocturnal analgesic response to morphine in mice. Physiol. Behav., 32: 261-264.

KEETON, W.T. (1971) Magnets interfere with pigeon homing. Proc. Natl Acad. Sci. (USA), 68: 102-106.

KHOLODOV, YU.A. (1964) Effects on the central nervous system. In: Barnothy, M.F., ed. Biological effects of magnetic fields, New York, London, Plenum Press, Vol. 1, pp. 196-200.

KHOLODOV, YU.A. (1966) [Effect of electromagnetic and magnetic fields on the central nervous system,] Moscow, Nauka (in Russian).

KHOLODOV, YU.A., ed. (1974) Influence of magnetic fields on biological objects, Springfield, Virginia, National Technical Information Service (NTIS Report No. JPRS 63038).

KHOLODOV, YU.A. (1981) [Peculiarities of the nervous system responses upon intensified magnetic fields.] Vopr. Kurort., 4: 5-9 (in Russian).

KHOLODOV, YU.A. (1982) [The brain in electromagnetic fields,] Moscow, Nauka, p. 120 (in Russian).

KHOLODOV, YU.A. & BERLIN, YU.V. (1984) [Sensory reactions in man during interactions with magnetic fields.] In: [Electromagnetic fields in the biosphere. II. Biological effects of electromagnetic fields,] Moscow, Nauka, pp. 83-89 (in Russian).

KHOLODOV, YU.A. & SHISHLO, M.A. (1980) [Electromagnetic fields in physiology,] Moscow, Nauka (in Russian).

KHOLODOV, YU.A., ALEXANDROVSKAYA, M.M., LUKJANOVA, S.N., & UDAROVA, N.S. (1969) Investigations of the reactions of mammalian brain to static magnetic fields. In: Barnothy, M., ed. Biological effects of magnetic fields, New York, London, Plenum Press, Vol. 2, pp. 215-225.

KIRSCHVINK, J.L. (1981) Ferromagnetic crystals (magnetite?) in human tissue. J. exp. Biol., 92: 333-335.

KIRSCHVINK, J.L. & LOWENSTAM, H.A. (1979) Mineralization and magnetization of chiton teeth: paleomagnetic, sedimentologic, and biologic implications of organic magnetite. Earth planet. Sci. Lett., 44: 193-204.

KLEIMENOVA, N.G. (1963) [Current concepts on the nature of high-frequency variations of the earth's electromagnetic field.] Izv. Av. SSR, 12: 1798-1813 (in Russian).

KLIMOVSKAYA, L.D. & MASLOVA, A.F. (1981) Stationary magnetic fields and reticular influences on adrenergic and cholinergic systems. Kosm. biol. aviakosm. Med., 15: 74-76.

KLIMOVSKAYA, L.D. & MASLOVA, A.F. (1983) Dynamics of changes in catecholamine and acetylcholine of blood during long-term exposure to high-induction stationary magnetic fields. Izv. Akad. Nauk SSR (Ser. Biol.), 4: 606-608.

KOLESOVA, N.I., VOLOSHINA, E.I., & UDINTSEV, N.A. (1978) Pathogenesis of insulin deficiency under the effect of an alternating magnetic field of industrial frequency. Patol. fiziol. Eksp., 6: 71-73.

KOLIN, A., BRILL, N.Q., & BROMBERG, P.J. (1959) Stimulation of irritable tissues by means of an alternating magnetic field. Proc. Soc. Exp. Biol. Med. (NY), 102: 251-253.

KOLODUB, F.A. & CHERNYSHEVA, O.N. (1980) Special features of carbohydrate-energy and nitrogen metabolism in the rat brain under the influence of magnetic fields of commercial frequency. Ukr. biokhim. Zh., 3: 299-303.

KOLODUB, F.A., CHERNYSHEVA, O.N., & EVTUSHENKO, G.I. (1981) [Mycocardial metabolic disturbances in rats exposed to alternating magnetic fields with different parameters.] Kardiologiia, 21: 82-85 (in Russian with English abstract).

KOMOLOVA, G.S., ERYGIN, G.D., VASIL'EVA, T.B., & EGOROV, I.A. (1972) Effect of a high-intensity constant magnetic field on enzymatic hydrolysis of nuclei acids. Dokl. Akad. Nauk. SSR, 204: 995-997.

KOPANOV, V.I. & SHAKULA, A.V. (1985) The effect of a hypogeomagnetic field on biological objects, Leningrad, Nauka, p. 73.

KOSSEL, F. (1982) Regulation and enforcement procedures, In: Suess, M.J., ed. Non-ionizing radiation protection, Copenhagen, World Health Organization, Regional Office for Europe (WHO Regional Publications, European Series No. 10).

KRAUSE, K., CREMER-BARTELS, G., KUCH, L.E., & WEITKAMPER, U. (1984) [The influence of low magnetic field variations on human night vision acuity.] Fortschr. Ophtalmol., 81: 183-185 (in German).

KRAUSE, N. (1986) Exposure of people to static and time variable magnetic fields in technology, medicine, research, and public life: dosimetric aspects. In: Bernhardt, J.H., ed. Biological effects of static and extremely low frequency magnetic fields, Munich, MMV Medizin Verlag, pp. 57-71.

KRONENBERG, S.S. & TENFORDE, T.S. (1979) Cell growth in a low-intensity, 60-Hz magnetic field, Berkeley, California, Lawrence Berkeley Laboratory (Report LBL-10050).

KREUGER, W.F., BRADLEY, J.W., GIAROLA, A.J., & DARUVALLA, S.R. (1972) Influence of low-level electric and magnetic fields on the growth of young chickens. ISA Trans., BM 72335: 183-186.

LABES, M.M. (1966) A possible explanation for the effect of magnetic fields on biological systems. Nature (Lond.), 211: 968.

LAMBDIN, J.D. (1978) A comparison of measurement techniques to determine electric fields and magnetic flux under EHV overhead power transmission lines, Las Vegas, Nevada, US Environmental Protection Agency (ORP/EAD 78-1).

LARKIN, R.P. & SUTHERLAND P.J. (1977) Migrating birds respond to Project Seafarer's electromagnetic field. Science, 195: 777-779.

LAWRENCE, A.F. & ADEY, W.R. (1982) Non-linear wave mechanisms in interactions between excitable tissue and electromagnetic fields. Neurol. Res., 4: 115-131.

LAWRENCE, A.F. & ADEY, W.R., ed. (1983) Non-linear electrodynamics in biological systems, New York, London, Plenum Press, p. 603.

LEASK, M.J.M. (1977) A physicochemical mechanism for magnetic field detection by migratory birds and homing pigeons. Nature (Lond.), 267: 144-145.

LEASK, M.J.M. (1978) Primitive models of magnetoreception. In: Schmidt-Koenig, K. & Keeton, W.T., ed. Animal migration, navigation, and homing, Berlin, Heidelberg, New York, Springer-Verlag, pp. 318-322.

LEE, J.M., BRUNKE, J.H., LEE, G.E., REINER, G.L., & SHON, F.L. (1982) Electrical and biological effects of transmission lines: a review, Portland, Oregon, US Department of Energy, Bonneville Power Administration.

LEE, P.H. & WEIS, J.J. (1980) Effects of magnetic fields on regeneration in fiddler crabs. Biol. Bull., 159-681.

LEVENGOOD, W.C. (1966) Cytogenetic variations induced with a magnetic probe. Nature (Lond.), 209: 1009-1013.

LEVENGOOD, W.C. (1967) Morphogenesis as influenced by locally-administered magnetic fields. Biophys. J., 7: 297-307.

LEVENGOOD, W.C. (1969) A new teratogenic agent applied to amphibian embryos. J. Embryol. exp. Morphol., 21: 23-31.

LIBERMAN, E.A., VAINTSVAIG, M.N., & TSOFINA, L.M. (1959) [The effect of a constant magnetic field on the excitation threshold of isolated frog nerve.] Biofizika, 4: 505-506 (in Russian with English abstract).

LIBOFF, A.R. (1985) Cyclotron resonance in membrane transport. In: Chiabrera, A., Nicolini, C., & Schwan, H.P., ed. Interactions beteen electromagnetic fields and cells, New York, London, Plenum Press, pp. 281-296.

LIBOFF, A.R., WILLIAMS, T., Jr, STRONG, N.N., & WISTAR, R., Jr (1984) Time-varying magnetic fields: effect on DNA synthesis. Science, 223: 818-820.

LIBOFF, A.R., THOMAS, J.R., & SCHROT, J. (1985) Magnetically induced behaviour modification in rats. In: Proceedings of the 7th Annual Meeting of the Bioelectromagnetics Society, San Francisco, California, 16-20 June, Gaithersburg, Maryland, Bioelectromagnetics Society, p. 22.

LIBOFF, R.L. (1980) Neuromagnetic thresholds. J. theor. Biol., 83: 427-436.

LIBURDY, R.P. & TENFORDE, T.S. (1986) Membrane responses to magnetic and electromagnetic fields. In: Maret, G., ed. Biophysical effects of steady magnetic fields, Berlin, Heidelberg, New York, Springer-Verlag.

LIBURDY, R.P., TENFORDE, T.S., & MAGIN, R.L. (1986) Magnetic field-induced drug permeability in liposome vesicles. Radiat. Res., 108(1): 102-111.

LIN, R.S., DISCHINGER, P.C., CONDE, J., & FARRELL, K.P. (1985) Occupational exposure to electromagnetic fields and the occurrence of brain tumors: an analysis of possible association. J. occup. Med., 27: 413-419.

LLNL (1985) Working in magnetic fields, Berkeley, University of California, Lawrence Livermore National Laboratory (Health and Safety Manual, Suppl. 26.12) (Revised by G. Miller, June 27).

LO, C.C., FUJITA, T.Y., & TENFORDE, T.S. (1986) A wide dynamic range portable 60 Hz magnetic dosimeter with data acquisition capabilities. IEEE Trans. Nucl. Sci., 33: 643-646.

LOVSUND, P., OBERG, P.A., & NILSSON, S.E.G. (1979) Influence on vision of extremely low frequency electromagnetic fields. Acta ophthalmol., 57: 812-821.

LOVSUND, P., NILSSON, S.E.G., OBERG, P.A., & REUTER, T. (1980a) Magneto phosphenes: a quantitative analysis of thresholds. Med. biol. Eng. Comput., 18: 326-334.

LOVSUND, P., OBERG, P.A., & NILSSON, S.E.G. (1980b) Magneto- and electrophosphenes: a comparative study. Med. biol. Eng. Comput., 18: 758-764.

LOVSUND, P., NILSSON, S.E.G., & OBERG, P.A. (1981) Influence on frog retina of alternating magnetic fields with special reference to ganglion cell activity. Med. biol. Eng. Comput., 19: 679-685.

LOVSUND, P., OBERG, P.A., & NILSSON, S.E.G. (1982) ELF magnetic fields in electrosteel and welding industries. Radio Sci., 17(5S): 35S-38S.

LOWENSTAM, H.A. (1962) Magnetite in denticle capping in recent chitons (Polyplacophora). Geol. Soc. Am. Bull., 73: 435-438.

LUBEN, R.A., CAIN, C.D., CHEN, M.C., ROSEN, D.M., & ADEY, W.R. (1982) Effects of electromagnetic stimuli on bone and bone cells in vitro: inhibition of responses to parathyroid hormone by low-energy low-frequency fields. Proc. Natl Acad. Sci. (USA), 79: 4180-4184.

LUNT, M.J. (1982) Magnetic and electric fields produced during pulsed magnetic field therapy for non-union of tibia. Med. biol. Eng. Comput., 20: 501-511.

LYLE, D.B., SCHECTER, P., ADEY, W.R., & LUNDAK, R.L. (1983) Suppression of T-lymphocyte cytotoxicity following exposure to sinusoidally amplitude modulated HF fields. Bioelectromagnetics, 4: 281-292.

LYU, B.N., YEFIMOV, M.L., KUL'SARTOV, V.K., & YAKUPOVA, R.M. (1978) [Movement of oxygen dissolved in liquid in a permanent magnetic field.] Biofizika, 23: 159-161 (in Russian with English summary).

MAASS, J.A. & ASA, M.M. (1970) Contactless nerve stimulation and signal detection by inductive transducer. IEEE Trans. Mag., 6: 322-326.

MCDOWALL, M.E. (1983) Leukaemia mortality in electrical workers in England and Wales. Lancet, 8318: 246.

MCDOWALL, M.E. (1986) Mortality of persons resident in the vicinity of electric transmission facilities. Br. J. Cancer, 53: 271-279.

MCLAUCHLAN, K.A. (1981) The effects of magnetic fields on chemical reactions. Sci. Prog. (Oxford), 67: 509-529.

MCLEOD, B.R. & LIBOFF, A.R. (1986) Dynamic characteristics of membrane ions in multifield configurations of low-frequency electromagnetic radiation. Bioelectromagnetics, 7: 177-189.

MCROBBIE, D. & FOSTER, M.A. (1984) Thresholds for biological effects of time-varying magnetic fields. Clin. phys. physiol. Meas., 5(2): 67-78.

MAFFEO, S., MILLER, M., & CARSTENSEN, E.L. (1984) Lack of effect of weak low frequency electromagnetic fields on chick embryogenesis. J. Anat., 139: 613-618.

MAGNUSSON, C.E. & STEVENS, H.C. (1911-12) Visual sensations caused by changes in the strength of a magnetic field. Am. J. Physiol., 29: 124-136.

MAHLUM, D.D. (1977) Biomagnetic effects: a consideration in fusion reactor development. Environ. Health Perspect., 20: 131-140.

MAHLUM, D.D., SIKOV, M.R., & DECKER, J.R. (1979) Dominant lethal studies in mice exposed to direct-current magnetic fields. In: Phillips, R.D., Gillis, M.F., Kaune, W.T., & Mahlum, D.D., ed. Biological effects of extremely low frequency electromagnetic fields, Springfield, Virginia, US Department of Energy, National Technical Information Service, pp. 474-484 (NTIS Report No. CONF-781016).

MALE, J.G., NORRIS, W.T., & WATTS, M.W. (1984) Human exposure to power frequency electric and magnetic fields. In:

Proceedings of the 23rd Hanford Life Sciences Symposium on Interaction of Biological Systems with Static and ELF Electric and Magnetic Fields, Richland, Washington, 2-4 October, Springfield, Virginia, National Technical Information Service.

MALING, J.E., WEISSBLUTH, M., & JACOBS, E.E. (1965) Enzyme substrate reactions in high magnetic fields. Biophys. J., 5: 767-776.

MALININ, G.I., GREGORY, W.D., MORELLI, L., SHARMA, V.K., & HOUCK, J.C. (1976) Evidence of morphological and physiological transformation of mammalian cells by strong magnetic fields. Science, 194: 844-846.

MANSFIELD, P. & MORRIS, P.G. (1982) NMR imaging in biomedicine. In: Waugh, J.S., ed. Biomagnetic effects, New York, Academic Press, Suppl. 2, pp. 247-252.

MANTELL, B. (1975) [Investigations into the effects on man of an alternating magnetic field at 50 Hz,] Freiburg, University of Freiburg (Ph.D. dissertation) (in German).

MARET, G. & DRANSFELD, K. (1977) Macromolecules and membranes in high magnetic fields. Physica, 86-88B: 1077-1083.

MARET, G. & DRANSFELD, K. (1985) Biomolecules and polymers in high steady magnetic fields. In: Herlach, F., ed. Applications of strong and ultrastrong magnetic fields. Topics in applied physics, Berlin, Heidelberg, New York, Springer-Verlag, Vol. 57, pp. 143-204.

MARET, G., SCHICKFUS, M.V., MAYER, A., & DRANSFELD, K. (1975) Orientation of nucleic acids in high magnetic fields. Phys. Rev. Lett., 35: 397-400.

MARKUZE, I.I., AMBARTSUMYAN, R.G., CHIBRIKIN, V.M., & PIRUZYAN, L.A. (1973) [Investigation of the PMP action on the alteration of the electrolyte concentration in the blood and organs of animals.] Izv. Akad. Nauk SSR (Ser. Biol.), 2: 281-283 (in Russian).

MARSAKOVA, N.V. (1983) [CNS role in the iodine content changes within organs and tissues under variable H-field conditions.] J. Kosm. biol. aviakosm. Med., 17: 90-92 (in Russian).

MARSH, J.L., AMSTRONG, T.J., JACOBSON, A.P., & SMITH, R.G. (1982) Health effects of occupational exposure to steady magnetic fields. Am. Ind. Hyg. Assoc. J., 43(6): 387-394.

MARTIN, H. & LINDAUER, M. (1977) The effects of the earth's magnetic field on gravity orientation in the honeybee (Apis mellifica). J. comp. Physiol., 122: 145.

MATHER, J.G. & BAKER, R.R. (1981) Magnetic sense of direction in woodmice for route-based navigation. Nature (Lond.), 291: 152-155.

MATHUR, D.V. (1984) Review Article. Biomedical implications of the relaxation behaviour of water related to NMR imaging. Br. J. Radiol., 57, 683: 955-976.

MEDVEDEV, M.A., URAZAEV, A.M., & KULAKOV, I.U.A. (1976) [Effect of a constant and low frequency magnetic field on the behavioural and autonomic responses of the human operator.] Zh. Vyssh. Nerv. Deiat., 26: 1131-1136 (in Russian with English abstract).

MELVILLE, D., PAUL, F., & ROATH, S. (1975) Direct magnetic separation of red cells from whole blood. Nature (Lond.), 255: 706.

MICHEL-BEYERLE, M.E., SCHEER, H., SEIDLITZ, H., TEMPUS, D., & HABERKORN, R. (1979) Time-resolved magnetic field effect on triplet formation in photosynthetic reaction center of Rhodopseudomonas sphaeroides R-26. FEBS Lett., 100: 9-12.

MILD, K.H., SANDSTROM, M., & LOVTRUP, S. (1981) Development of Xenopus laevis embryos in a static magnetic field. Bioelectromagnetics, 2: 199-201.

MILHAM, S., Jr. (1979) Mortality in aluminium reduction plant workers. J. occup. Med., 21: 475-480.

MILHAM, S., Jr. (1982) Mortality from leukaemia in workers exposed to electrical and magnetic fields. New Engl. J. Med., 307(4): 249.

MILHAM, S., Jr (1985a) Silent keys: leukaemia mortality in amateur radio operators. Lancet, 1(8432): 812.

MILHAM, S., Jr (1985b) Mortality in workers exposed to electro- magnetic fields. Environ. Health Perspect., 62: 297-300.

MILLER, M.W. (1980) Re: Electrical wiring configurations and childhood cancer. Am. J. Epidemiol., 112: 165-167.

MITBREIT, I.M. & MANYACHIN, V.D. (1984) Influence of magnetic fields on the repair of bone, Moscow, Nauka, pp. 292-296.

MITTLER, S. (1971) Failure of magnetism to influence production of X-ray induced sex-linked recessive lethals. Mutat. Res., 13: 287-288.

MIZUSHIMA, Y., AKAOKA, I., & NISHIDA, Y. (1975) Effects of magnetic field on inflammation. Experientia (Basel), 21: 1411-1412.

MOLIN, YU.N., SAGDEEV, R.Z., & SALIKHOV, K.M. (1979) Effects of magnetic field on radical reactions in solution. In: Vol'Pin, M.E., ed. Soviet scientific review section B, Vol. 1, pp. 1-67 (Chemistry Review Series).

MONTGOMERY, D.J. & SMITH, A.E. (1963) A search for biological effects of magnetic fields. Biomed. Sci. Instrum., 1: 123.

MOSS, A.J. & CARSTENSEN, E. (1985) Evaluation of the effects of electric fields on implanted cardiac pacemakers, Palo Alto, California, Electric Power Research Institute, pp. 679-6.

MULAY, I.L. & MULAY, L.N. (1961) Effect of a magnetic field on Sarcoma 37 ascites tumour cells. Nature (Lond.), 190: 1019.

MULAY, I.L. & MULAY, L.N. (1964) Effects on Drosophila melanogaster and S-37 tumour cells: postulates for magnetic field interactions. In: Barnothy, M.F., ed. Biological effects of magnetic fields, New York, London, Plenum Press, Vol. 1, pp. 146-169.

MULLER, K., HABERDITZL, W., & PRITZE, B. (1971) Examination of the influence of magnetic fields on chemical reactions. Z. Phys. Chem., 248: 185-192.

MURAYAMA, M. (1965) Orientation of sickle cell erythrocytes in a magnetic field. Nature (Lond.), 206: 420-422.

MURRAY, J.C. & FARNDALE, R.W. (1985) Modulation of collagen production in cultured fibroblasts by a low-frequency, pulsed magnetic field. Biochim. Biophys. Acta, 838: 98-105.

MYERS, A., CARTWRIGHT, J.A, BONNELL, J.A., MALE, J.C., & CARTWRIGHT, S.C. (1985) Overhead power lines and childhood cancer. In: Proceedings of the International Conference on

Electric and Magnetic Fields in Medicine and Biology, London, October 1985.

NAHAS, G.G., BOCCALON, H., BERRYER, P., & WAGNER, B. (1975) Effects in rodents of a 1-month exposure to magnetic fields (200-1200 Gauss). Aviat. Space environ. Med., 46: 1161-1163.

NAKAGAWA, M. (1979) Effects of magnetic fields on fertility, general reproductive performance and growth of mice. Jpn J. Hyg., 34: 488-495.

NAKAGAWA, M., MUROYA, H., MATSUDA, Y., & TSUKAMOTO, H. (1980) [Effects of static magnetic field on some lipid and protein metabolic processes of rabbit.] J. Transp. Med., 34: 376-384 (in Japanese).

NAKHIL'NITSKAYA, Z.N., MASTRYUKOVA, V.M., ANDRIANOV, L.A., & BOROUKINA, A.G. (1978) [Organism response to "zero magnetic fields" impact.] Kosm. biol. aviakosm. Med., 12(2): 74-76 (in Russian).

NATH, R., SCHULZ, R.J., & BONGIORNI, P. (1980) Response to mammalian cells irradiated with 30MV X-rays in the presence of a uniform 20-kilogauss magnetic field. Int. J. Radiat. Biol., 38: 285-292.

NAZAROVA, N.M., LIVSHITZ, V.A., ANZIN, V.B., VESELAGO, V.G., & KUZENTSOV, A.N. (1982) [Hydrolysis of globular proteins by trypsin in a strong magnetic field.] Biofizika, 27: 720 (in Russian with English abstract).

NEUGEBAUER, D.-CH., BLAUROCK, A.E., & WORCESTER, D.L. (1977) Magnetic orientation of purple membranes demonstrated by optical measurements and neutron scattering. Fed. Eur. Biomed. Soc. Lett., 78: 31-35.

NEURATH, P.W. (1968) High gradient magnetic field inhibits embryonic development of frogs. Nature (Lond.), 219: 1358-1359.

NEW, P.F.J., ROSEN, B.R., BRADY, T.J., BUONANNO, F.S., KISTLER, J.P., BURT, C.T., HINSHAW, W.S., NEWHOUSE, J.H., POHOST, G.M., & TAVERAS, J.M. (1983) Potential hazards and artifacts of ferromagnetic and non-ferromagnetic surgical and dental materials and devices in magnetic resonance imaging. Radiology, 147: 139-148.

NORTON, L.A. (1982) Effects of a pulsed electromagnetic field on a mixed chondroblastic tissue culture. Clin. orthop. relat. Res., 167: 280-290.

NRPB (1981) Exposure to nuclear magnetic resonance clinical imaging. Radiography, 47(563): 258-260.

NRPB (1983) Ad hoc Advisory Group on Nuclear Magnetic Resonance Clinical Imaging: Revised guidance on acceptable limits of exposure during nuclear resonance clinical imaging. Br. J. Radiol., 56: 974-977.

NRPB (1984) Advice on acceptable limits of exposure to nuclear magnetic resonance clinical imaging, Chilton, Didcot, Oxon, H.M. Stationery Office.

OBERG, P.A. (1973) Magnetic stimulation of nerve tissue. Med. biol. Eng., 11: 55-64.

ODINTSOV, Y.N. (1965) [The effect of a magnetic field on the natural resistance of white mice to Listeria infection.] Tr. Tomsk. Vaktsyn. Syvorotok, 16: 234-238 (in Russian with English abstract).

OGRODNIK, A., KRUGER, H.W., ORTHUBER, H., HABERKORN, R., MICHEL-BEYERLE, M.E., & SCHEER, H. (1982) Recombination dynamics in bacterial photosynthetic reaction centers. Biophys. J., 39: 91-99.

OLCESE, J. & REUSS, S. (1986) Magnetic field effects on pineal gland melatonin synthesis: comparative studies on albino and pigmented rodents. Brain Res., 369: 365-368.

OLCESE, J., REUSS, S., & VOLLRATH, L. (1985) Evidence for the involvement of the visual system in mediating magnetic field effects on pineal melatonin synthesis in the rat. Brain Res., 333: 382-384.

OLIN, R., VAGERO, D., & AHLBOM, A. (1985) Mortality experience of electrical engineers. Br. J. ind. Med., 42: 211-212.

OSSENKOPP, K.-P. & SHAPIRO, J. (1972) Effects of prenatal exposure to a 0.5 Hz low-intensity rotating magnetic field on White Peking ducklings. Am. Zool., 12: 650.

OSSENKOPP, K.-P., KOLTEK, W.T., & PERSINGER, M.A. (1972) Prenatal exposure to an extremely low frequency-low intensity rotating magnetic field and increase in thyroid and testicle weight in rats. Dev. Psychobiol., 5: 275-285.

PAPI, F., MESCHINI, E., & BALDACCINI, N.E. (1983) Homing behaviour of pigeons released after having been placed in alternating magnetic fields. Comp. Biochem. Biophys., 76A: 673-682

PAUL, F., ROATH, S., & MELVILLE, D. (1978) Differential blood cell separation using a high gradient magnetic field. Br. J. Haematol., 38: 273-280.

PAUTRIZEL, R., PRIORE, A., BERLUREAU, F., & PAUTRIZEL, A.N. (1969) Stimulation, par des moyens physiques, des défenses de la souris et du rat contre la trypanosomose expérimentale. C. R. Acad. Sci. (Paris) Ser. D., 268: 1889-1892.

PAVLICEK, W., GEISINGER, M., CASTLE, L., BORKOWSKI, G.P., MEANEY, T.F., BREAM, B.L., & GALLAGHER, J.H. (1983) The effects of nuclear magnetic resonance on patients with cardiac pacemakers. Radiology, 147: 149-153.

PEARCE, N.E., SHEPPARD, R.A., HOWARD, J.K., FRASER, J., & LILLEY, B.M. (1985) Leukaemia in electrical workers in New Zealand. Lancet, 1(8432): 811-812.

PELYHE, I., MESZAROS, I., & SARVARI, E. (1973) Effects of static magnetic field on the establishment of conditioned electrodefensive reflex in the rat. Acta physiol. Acad. Sci. Hung., 43: 125-132.

PERAKIS, N. (1947) Sur la croissance des cultures de fibroblastes dans un champ magnétique. Acta anat., 4: 225.

PEREIRA, M.R., NUTINI, L.G., FARDON, J.C., & COOK, E.S. (1967) Cellular respiration in intermittent magnetic fields. Proc. Soc. Exp. Biol. Med., 124: 573-576.

PERRY, F.S., REICHMANIS, M., MARINO, A.A., & BECKER, R.O. (1981) Environmental power frequency magnetic fields and suicide. Health Phys., 41: 267-277.

PERSINGER, M.A. (1969) Open-field behaviour in rats exposed prenatally to a low intensity-low frequency, rotating magnetic field. Dev. Psychobiol., 2: 168-171.

PERSINGER, M.A. & CODERRE, D.J. (1978) Thymus mast cell numbers following perinatal and adult exposures to low intensity 0.5 Hz magnetic fields. Int. J. Biometeorol., 22: 123-128.

PERSINGER, M.A. & FOSTER, W.S. (1970) ELF rotating magnetic fields: prenatal exposure and adult behaviour. Arch. Met. Geoph. Biokl. (Ser. B), 18: 363-369.

PERSINGER, M.A. & PEAR, J.J. (1972) Prenatal exposure to an ELF-rotating magnetic field and subsequent increase in conditioned suppression. Dev. Psychobiol., 5: 269-274.

PERSINGER, M.A., OSSENKOPP, K.-P., & GLAVIN, G.B. (1972) Behavioural changes in adult rats exposed to ELF magnetic fields. Int. J. Biometeorol., 16: 155-162.

PERSINGER, M.A., LAFRENIERE, G.F., & CARREY, N.J. (1978) Thyroid morphology and wet organ weight changes in rats exposed to different low intensity 0.5 Hz magnetic fields and pre-experimental caging conditions. Int. J. Biometeorol., 22: 67-73.

PERSSON, B.R.R. & STAHLBERG, F. (1984) Potential health hazards and safety aspects of clinical NMR examinations, Lund, Radiation Physics Departments.

PHILLIPS, J.L., WINTERS, W.D., & RUTLEDGE, L. (1986a) In vitro exposure to electromagnetic fields: changes in tumour cell properties. Int. J. Radiat. Biol., 49(3): 463-469.

PHILLIPS, J.L., RUTLEDGE, L., & WINTERS, W.D. (1986b) Transferrin binding to two human colon carcinoma cell lines: characterization and effects of 60-Hz electromagnetic fields. Cancer Res., 46: 239-244.

PILLA, A.A. (1979) Electrochemical information transfer and its possible role in the control of cell function. In: Brighton, C.T., Black, H., & Pollack, S.R., ed. Electrical properties of bone and cartilage, New York, London, Grune and Stratton, pp. 455-489.

PILLA, A.A., SECHAUD, P., & MCLEOD, B. (1983) Electrochemical and electrical aspects of low-frequency current induction in biological systems. J. Biol. Phys., 11: 51-57.

PIRUSYAN, L.A. & KUZNETSOV, A.N. (1983) [The study of mechanisms of constant and low frequency magnetic fields influence on the biological systems.] In: [Proceedings of the USSR Academy of Sciences, November - December,] Pushchino, USSR Academy of Sciences (Biological Series No. 6) (in Russian).

POLK, C. (1974) Sources, propagation, amplitude and temporal variation of extremely low frequency (0-100Hz) electromagnetic fields: In: Llaurado, J.G., Jances, A., & Battocletti, J.H., ed. Biological and clinical effects of low frequency magnetic and electric fields, Springfield, Illinois, Charles C. Thomas, pp. 21-48.

POLK, C. (1984) Motion of counterions on a cylindrical cell surface: a possible mechanism for the action of low-frequency, low-intensity magnetic fields which displays unsuspected frequency dependence. In: Proceedings of the 23rd Hanford Life Sciences Symposium on Interaction of Biological Systems with Static and ELF Electric and Magnetic Fields, Richland, Washington, 2-4 October, Springfield, Virginia, National Technical Information Service.

POLSON, M.J.R., BARKER, A.T., & FREESTON, I.L. (1982) Stimulation of nerve trunks with time-varying magnetic fields. Med. biol. Eng. Comput., 20: 243-244.

PRESTI, D. & PETTIGREW, J.D. (1980) Ferromagnetic coupling to muscle receptors as a basis for geomagnetic field sensitivity in animals. Nature (Lond.), 285: 99-101.

RABINOVITCH, B., MALING, J.E., & WEISSBLUTH, M. (1967a) Enzymesubstrate reactions in very high magnetic fields. I. Biophys. J., 7: 187-204.

RABINOVITCH, B., MALING, J.E., & WEISSBLUTH, M. (1967b) Enzymesubstrate reactions in very high magnetic fields. II. Biophys. J., 7: 319-327.

RABINOVITCH, E.Z., TARAN, Y.P., USACHEVA, M.D., EPSHTEYN, T.M., & KUZNETSOV, A.N. (1983) [The effect of constant magnetic field on human skin respiration in reparative and destructive processes.] Biofizika, 28: 693-696 (in Russian with English abstract).

RAMIREZ, E., MONTEAGUDO, J.L., GARCIA-GARCIA, M., & DELGADO, J.M.R. (1983) Oviposition and development of Drosophila modified by magnetic fields. Bioelectromagnetics, 4: 315-326.

RAMON, C., AYAZ, M., & STREETER, D.D., Jr (1981) Inhibition of growth rate of Escherichia coli induced by extremely low-frequency weak magnetic fields. Bioelectromagnetics, 2: 285-289.

RATNER, S.C. (1976) Kinetic movements in magnetic fields of chitons with ferromagnetic structures. Behav. Biol., 17: 573-578.

RAYBOURN, M.S. (1983) The effects of direct-current magnetic fields on turtle retinas in vitro. Science, 220: 715-717.

RENO, V.R. (1969) Conduction velocity in nerve exposed to a high magnetic field, Pensacola, Florida, Naval Aerospace Medical Institute (Report No. AD-699-171).

RENO, V.R. & NUTINI, L.G. (1963) Effect of magnetic field on tissue respiration. Nature (Lond.), 198: 204-205.

RENO, V.R. & NUTINI, L.G. (1964) Tissue respiration. In: Barnothy, M.F., ed. Biological effects of magnetic fields, New York, London, Plenum Press, Vol. 1, pp. 211-217.

REPACHOLI, M.H. (1983a) Differentiation between biological effects and health hazards: scaling from animals. In: Grandolfo, M., Michaelson, S.M., & Rindi, A., ed. Biological effects and dosimetry of non-ionizing radiation, New York, London, Plenum Press, pp. 531-548.

REPACHOLI, M.H. (1983b) Development of standards-assessment of health hazards and other factors. In: Grandolfo, M., Michaelson, S.M., & Rindi, A., ed. Biological effects and dosimetry of non-ionizing radiation, New York, London, Plenum Press, pp. 611-625.

REPACHOLI, M.H. (1985a) Video display terminals - should operators be concerned? Australas. phys. eng. Sci. Med., 8(2): 51-61.

REPACHOLI, M.H. (1985b) Health risk assessment of static and ELF electric and magnetic fields (0-300Hz). In: Grandolfo, M., Michaelson, S., & Rindi, A., ed. Biological effects and dosimetry of static and ELF electromagnetic fields, New York, London, Plenum Press.

REPACHOLI, M.H. (1986) Limits of human exposure to magnetic fields. In: Bernhardt, J.H., ed. Biological effects of static and extremely low frequency magnetic fields, Munich, MMV Medizin Verlag, pp. 171-180.

REUSS, S. & OLCESE, J. (1986) Magnetic field effects on the rat pineal gland: role of retinal activation by light. Neurosci. Lett., 64: 97-101.

REUSS, S., SEMM, P., & VOLLRATH, L. (1983) Different types of magnetically sensitive cells in the rat pineal gland. Neurosci. Lett., 40: 23-26.

REUSS, S., OLCESE, J., VOLLRATH, L., SKALEJ, M., & MEVES, M. (1985) Lack of effect of NMR-strength magnetic fields on rat pineal melatonin synthesis. IRCS Med. Sci., 13: 471.

RIESEN, W.H., ARANYI, C., KYLE, J.L., VALENTINO, A.R., & MILLER, D.A. (1971) A pilot study of the interaction of extremely low frequency electromagnetic fields with brain organelles, Bethesda, Maryland, Naval Medical Research Development Command (Compilation of Navy-Sponsored ELF Biomedical and Ecological Research Reports Vol. 1) (Technical Memo No. 3) (IITR Project No. E6185).

ROCKETTE, H.E. & ARENA, V.C. (1983) Mortality studies of aluminum reduction plant workers: potroom and carbon department. J. occup. Med., 25: 549-557.

ROCKWELL, S. (1977) Influence of a 1400 Gauss magnetic field on the radiosensitivity and recovery of EMT6 cells in vitro. Int. J. Radiat. Biol., 31: 153-160.

RODVALL, Y., FEYCHTING, M., & AHLBOM, A. (1985) [Investigation of morbidity from cancers in the Alfta area,] Stockholm, National Institute of Environmental Medicine (in Swedish).

ROOTS, R.J., KRAFT, G.H., FARINATO, R.S., & TENFORDE, T.S. (1982) Electrophoretic and electrooptical studies on the conformation and susceptibility to psoralen crosslinking of magnetically oriented DNA, Berkeley, University of California, Lawrence Berkeley Laboratory (Report No. LBL-13601).

ROSENBLATT, C., TORRES DE ARAUJO, F.F., & FRANKEL, R.B. (1982a) Light scattering determination of magnetic moments of magnetotactic bacteria. J. appl. Phys., 53: 2727.

ROSENBLATT, C., TORRES DE ARAUJO, F.F., & FRANKEL, R.B. (1982b) Birefringence determination of magnetic moments of magnetostatic bacteria. Biophys. J., 40: 83-85.

ROSCHIN, V.A. (1985) [Assessment of local magnetic field effects on human volunteers.] Gig. Tr. Prof. Zabol., 7: 33-36 (in Russian).

ROY, C.R., JOYNER, K.H., GIES, H.P., & BANGAY, M.J. (1984) Measurement of electromagnetic radiation emitted from visual display terminals (VDTs). Radiat. Prot. Aust., 2(1): 26-30.

RUSSELL, D.R. & HENDRICK, H.G. (1969) Preference of mice to consume food and water in an environment of high magnetic field. In: Barnothy, M.F., ed. Biological effects of magnetic fields, New York, London, Plenum Press, Vol. 2, pp. 233-239.

RUSYAYEV, V.F. (1979) [Effects of magnetism on thrombocyte coagulation.] Prob. Gematol. Pereliv. Krovi., 2: 19-23 (in Russian).

RYABY, J.T., GOODMAN, R., HENDERSON, A.S., & BASSETT, C.A.L. (1983) Electromagnetic field effects on cellular biosynthetic processes. Trans. BRAGS, 111: 25.

SAKHAROVA, S.A., RYZHOV, A.I., & UDINTSEV, N.A. (1977) [Reaction of central and peripheral mediator elements of the sympathoadrenal system to single exposures to alternating magnetic fields.] Dokl. Vyssh. Shk. Biol. Nauki, 9: 35-39 (in Russian).

SAKHAROVA, S.A., RYZHOV, A.I., & UDINTSEV, N.A. (1981) Mechanism of the sympathoadrenal system's response to the one time action of a variable magnetic field. Kosm. biol. aviakosm. Med., 15: 52-56.

SANDER, R., BRINKMANN, J., & KUHNE, B. (1982) Laboratory studies on animals and human beings exposed to 50 Hz electric and magnetic fields. In: Proceedings of the International Congress on Large High Voltage Electric Systems, Paris, 1-9 September, Paris, CIGRE (CIGRE Paper 36-01).

SAUNDERS, R.D. & CASS, A. (1983) Magnetic field interactions with living systems, Didcot, Berkshire, National Radiological Protection Board (Report M96).

SAUNDERS, R.D. & SMITH, H. (1984) Safety aspects of NMR clinical imaging. Br. med. Bull., 40(2): 148-154.

SCHMITT, F.O., SCHNEIDER, D.M., & CROTHERS, D.M., ed. (1975) Functional linkage in biomolecular systems, New York, Raven Press.

SCHULTEN, K. (1982) Magnetic field effects in chemistry and biology. Adv. solid-state Phys., 22: 61-83.

SCHULTEN, K. (1986) Magnetic fields effects on radical pair processes in chemistry and biology. In: Bernhardt, J.H., ed. Biological effects of static and ELF magnetic fields, Munich, MMV Medizin-Verlag, pp. 133-140.

SCHULTEN, K., SWENBERG, CH.E., & WALLER, A. (1978) A biomagnetic sensory mechanism based on magnetic field modulated coherent electron spin motion. Ztg. Phys. Chem., 111: 1-5.

SCHWARTZ, J.L. (1978) Influence of a constant magnetic field on nervous tissues. I. Nerve conduction velocity studies. IEEE Trans. Biomed. Eng., 25: 467-473.

SCHWARTZ, J.L. (1979) Influence of a constant magnetic field on nervous tissues. II. Voltage-clamp studies. IEEE Trans. Biomed. Eng., 26: 238-243.

SCOTT-WALTON, B., CLARK, K.M., HOLT, B.R., JONES, D.C., KAPLAN, S.D., KREBS, J.S., POLSON, P., SHEPHERD, R.A., & YOUNG, J.R. (1979) Potential environmental effects of 765-kV transmission lines: views before the New York State Public Service Commission. Cases 26529 and 26559, 1976-1978, Springfield, Virginia, US Department of Energy, National Technical Information Service (NTIS Report No. DOE/EV-0056).

SEIDEL, D., KNOLL, M., & EICHMEIER, J. (1968) [Excitation of subjective light flashes in man (phosphenes) by a sinusoidal magnetic field.] Pflüger Arch. gesamte Physiol., 299: 11-18 (in German).

SEMM, P. (1983) Neurobiological investigations on the magnetic sensitivity of the pineal gland in rodents and pigeons. Comp. Biochem. Physiol., 76A: 683-689.

SEMM, P. (1986) Sensitivity to natural magnetic fields in the central nervous system of pigeons. In: Bernhardt, J.H., ed. Biological effects of static and extremely low frequency magnetic fields, Munich, MMV Medizin-Verlag, pp. 96-101.

SEMM, P., SCHNEIDER, T., & VOLLRATH, L. (1980) Effects of an earth-strength magnetic field on electrical activity of pineal cells. Nature (Lond.), 288: 607-608.

SEMM, P., SCHNEIDER, T., VOLLRATH, L., & WILTSCHKO, W. (1982) Magnetic sensitive pineal cells in pigeons. In: Papi, F. & Wallraff, H.G., ed. Avian navigation, Berlin, Heidelberg, New York, Springer-Verlag, pp. 329-337.

SEMM, P., NOHR, D., DEMAINE, C., & WILTSCHKO, W. (1984) Neural basis of the magnetic compass: interactions of visual, magnetic and vestibular inputs in the pigeons brain. J. comp. Physiol. A., 155: 283-288.

SHEPPARD, A.R. (1983) Results of exposure of Aplysia pacemaker neurons to ELF/60 Hz and DC magnetic fields. Proceedings of the 5th Annual Meeting of the Bioelectromagnetics Society, Boulder, Colorado, 12-16 June, Gaithersburg, Maryland, Bioelectromagnetics Society, p. 25.

SHEPPARD, A.R. (1985) Cellular studies of effects of ELF electric and magnetic fields. In: Biological and human health effects of extremely low frequency electromagnetic fields, Arlington, Virginia, American Institute of Biological Sciences, pp. 129-184.

SHEPPARD, A.R. & EISENBUD, M. (1977) Biological effects of electric and magnetic fields of extremely low frequency, New York, New York University Press.

SHISLO, M.A. (1974) Influence of magnetic fields on enzymes, tissue respiration, and some aspects of metabolism in an intact organism, In: Kholodov, Y.A., ed. Influence of magnetic fields on miological objects, Springfield, Virginia, National Technical Information Service, pp. 20-25 (NTIS Report No. JPRS 63038).

SHOBER, A., YANK, M., & FISCHER, G. (1982) Electrolyte changes in the white mouse under the influence of a weak magnetic field. Zbl. Bakteriol. Hyg., B176: 305-315.

SHUST, I.V., GALANTYUK, S.I., & CHERETYANKO, YU.V. (1980) Particular features of conditioned electrodefensive reflex in white rats on a background of constant magnetic field. Fiziol. Zh., 26: 264-268.

SIDJAKIN, V.G. (in press) [The influence of global ecological factors on the nervous system,] Kiev, Naukova Dumka (in Russian).

SIKOV, M.R., MAHLUM, D.D., MONTGOMERY, L.D., & DECKER, J.R. (1979) Development of mice after intrauterine exposure to direct-current magnetic fields. In: Phillips, R.D., Gillis, M.F., Kaune, W.T., & Mahlum, D.D., ed. Biological effects of extremely low frequency electromagnetic fields, Springfield, Virginia, US Department of Energy, National Technical Information Service, pp. 462-473 (NTIS Report No. CONF-791016).

SILNY, J. (1981) Influence of low-frequency magnetic field (LMF) on the organism. In: Proceedings of the 4th Symposium on Electromagnetic Compatibility, Zurich, 10-12 March, pp. 175-180 (Paper 33.G2).

SILNY, J. (1984) Changes in VEP caused by strong magnetic fields. In: Nodar, R.H. & Barber, C., ed. Evoked Potentials II: The Second International Evoked Potentials Symposium, Boston, Butterworth, pp. 272-279.

SILNY, J. (1986) The influence threshold of the time varying magnetic field in the human organism. In: Bernhardt, J.H., ed. Biological effects of static and extremely low frequency magnetic fields, Munich, MMV Medizin-Verlag, pp. 105-112.

SINCLAIR, W.K. (1981) The scientific basis for risk quantification. In: Proceedings of 6th Annual Meeting, 2-3 April, 1980, Washington DC, National Council on Radiation Protection and Measurements, pp. 3-33.

SLINEY, D.H. (1986) Does the basis for a standard really exist? In: Bernhardt, J.H., ed. Biological effects of static and extremely low frequency magnetic fields, Munich, MMV Medizin-Verlag, pp. 181-183.

SMIRNOVA, N.P. (1982) [Behaviour of rats in "open field" following magnetic field exposure.] J. Vyssh. Nervnoj Deyat., 32(1): 72 (in Russian).

SMITH, C.W. (1982) Comments on the paper "Environmental power-frequency magnetic fields and suicide." Health Phys., 43: 439-441.

SMITH, R.F. & JUSTESEN, D.R. (1977) Effects of a 60 Hz magnetic field on activity levels of mice. Radio Sci., 12(6S): 279-285.

SOLDATOVA, L.P. (1982) Sequence of pathomorphological reactions to the effect of alternating magnetic fields. Arkh. Anat. Gistol. Embriol., 83: 12-15.

SOUTHERN, W.E. (1975) Orientation of gull chicks exposed to Project Sanguine's electromagnetic field. Science, 189: 143-145.

SPERBER, D., OLDENBOURG, E., & DRANSFELD, K. (1984) Magnetic field induced temperature change in mice. Naturwissenschaften, 71: 100-101.

STANFORD LINEAR ACCELERATOR CENTER (1970) Limits on human exposure to static magnetic fields, Palo Alto, California, Stanford Linear Accelerator Center.

STERN, F.B., WAXWEILER, R.A., BEAUMONT, J.J., LEE, S.T., RINSKY, R.A., ZUMWALDE, R.D., HALPERIN, W.E., BIERBAUM, P.J., LANDRIGAN, P.J., & MURRAY, W.E. (1986) A case-control study of leukaemia at a naval nuclear shipyard. Am. J. Epidemiol., 123: 980-992.

ST. LORANT, S.J. (1977) Biomagnetism: a review, Palo Alto, California, Stanford Linear Accelerator Center (Publication No. 1984).

STRAND, J.A., ABERNATHY, C.S., SKALSKI, J.R., & GENOWAY, R.G. (1983) Effects of magnetic field exposure on fertilization success in rainbow trout. Salmo gairdneri. Bioelectromagnetics, 4: 295-301.

STRZHIZHOVSKY, A.D. & MASTRYUKOVA, V.M. (1983) Effect of high-intensity constant magnetic fields on spermatogenesis of mammals. Izv. Akad. Nauk SSR (Ser. Biol.), 3: 473-475.

STRZHIZHOVSKY, A.D., GALAKTIONOVA, G.V., & CHEREMNYKH, P.A. (1980) [Tissue specificity of mitotic activity changes under the influence of strong magnetic fields.] Tsitologiya, 12(2): 205-208 (in Russian).

STUCHLY, M.A. (1986) Exposure to static and time-varying magnetic fields in industry research and public life: dosimetric aspects. In: Bernhardt, J.H., ed. Biological effects of static and extremely low frequency magnetic fields, Munich, MMV Medizin Verlag, pp. 39-56.

STUCHLY, M.A. & LECUYER, D.W. (1985) Induction heating and operator exposure to electromagnetic fields. Health Phys., 49: 693-700.

STUCHLY, M.A., LECUYER, D.W., & MANN, R.D. (1983) Extremely low frequency electromagnetic emissions from video display terminals and other devices. Health Phys., 45: 713-722.

SWERDLOW, A.J. (1983) Epidemiology of eye cancer in adults in England and Wales, 1962-1977. Am. J. Epidemiol., 118(2): 294-300.

SWICORD, M.L. (1985) Possible biophysical mechanisms of electromagnetic interactions with biological systems. Am. Inst. Biol. Res.: 61-78.

TABRAH, F.L., GUERNSEY, D.L., CHOU, S.-C., & BATKIN, S. (1978) Effect of alternating magnetic fields (60-100 Gauss, 60 Hz) on Tetrahymena pyriformis. TIT life Sci., 8: 73-77.

TARAKHOVSKY, M.L., SAMBROSKAYA, E.P., MEDVEDEV, B.M., ZADEROZHNAYA, T.D., OKHRONCHUK, B.V., & LIKHTENSHTEIN, E.M. (1971) Effects of constant and alternating magnetic fields on some indices of physiological functions and metabolic processes in albino rats. Fiziol. Zh. Kiev Akad. Nauk Ukr. (RSR), 17: 452-459.

TEGENKAMP, T.R. (1969) Mutagenic effects of magnetic fields on Drosophila melanogaster. In: Barnothy, M.F., ed. Biological effects of magnetic fields, New York, London, Plenum Press, Vol. 2, pp. 189-206.

TELL, R.A. (1983) Instrumentation for measurement of radiofrequency electromagnetic fields: equipment, calibrations, and selected applications. In: Grandolfo, M., Michaelson, S., & Rindi, A., ed. Biological effects and dosimetry of non-ionizing radiation: radiofrequency and microwave energies, New York, London, Plenum Press, pp. 95-162.

TEMUR'YANTS, M.A., YEVSTAF'YEVA, YE.V., & MAKEYEV, V.B. (1985) [Adaptation of lipid metabolism in rats with restricted mobility as a result of exposure to a variable infra-low frequency magnetic field.] Biofizika, 30: 313-316 (in Russian with English summary).

TENFORDE, T.S., ed. (1979) Magnetic field effects on biological systems, New York, London, Plenum Press.

TENFORDE, T.S. (1984) Interaction of stationary magnetic fields with the cardiovascular system, Berkeley, University of California, Lawrence Berkeley Laboratory (Report No. LBL-18329).

TENFORDE, T.S. (1985a) Mechanisms for biological effects of magnetic fields. In: Grandolfo, M., Michaelson, S.M., & Rindi, A., ed. Biological effects and dosimetry of static and ELF electromagnetic fields, New York, London, Plenum Press, pp. 71-92.

TENFORDE, T.S. (1985b) Biological effects of stationary magnetic fields. In: Grandolfo, M., Michaelson, S.M., & Rindi, A., ed. Biological effects and dosimetry of static and ELF electromagnetic fields, New York, London, Plenum Press, pp. 93-127.

TENFORDE, T.S. (1985c) Biological effects of ELF magnetic fields. In: Biological and human health effects of extremely low frequency electromagnetic fields, Arlington, Virginia, American Institute of Biological Sciences, pp. 79-127.

TENFORDE, T.S. (1986a) Interaction of ELF magnetic fields with living matter. In: Polk, C. & Postow, E., ed. Handbook of biological effects of electromagnetic fields, Boca Raton, Florida, CRC press, pp. 197-225.

TENFORDE, T.S. (1986b) Magnetic field applications in modern technology and medicine. In: Bernhardt, J.H., ed. Biological effects of static and ELF magnetic fields, Munich, MMV Medizin-Verlag, pp. 23-35.

TENFORDE, T.S. (1986c) Thermoregulation in rodents exposed to high-intensity stationary magnetic fields. Bioelectromagnetics, 7: 341-346.

TENFORDE, T.S. (1986d) Biological effects of extremely low frequency magnetic fields. In: Feero, W.E., ed. Panel Session on Biological Effects of Power-Frequency Electric and Magnetic Fields, Piscataway, New Jersey, Institute of Electrical and Electronic Engineering, pp. 21-40 (IEEE Publication No. 86THO139-6-PWR).

TENFORDE, T.S. & BUDINGER, T.F. (1986) Biological effects and physical safety aspects of NMR imaging and in vivo spectroscopy. In: Thomas, S.R., ed. NMR in medicine: instrumentation and clinical applications, New York, American Association of Physicists in Medicine (Monograph Series).

TENFORDE, T.S. & KAUNE, W.T. (in press) Interaction of extremely low frequency electric and magnetic fields with living systems. Health Phys..

TENFORDE, T.S. & SHIFRINE, M. (1984) Assessment of the immune responsiveness of mice exposed to a 1.5 Tesla stationary magnetic field. Bioelectromagnetics, 5: 443-446.

TENFORDE, T.S., GAFFEY, C.T., MOYER, B.R., & BUDINGER, T.F. (1983) Cardiovascular alterations in Macaca monkeys exposed to stationary magnetic fields: experimental observations and theoretical analysis, Bioelectromagnetics, 4: 1-9.

TENFORDE, T.S., GAFFEY, C.T., & RAYBOURN, M.S. (1985) Influence of stationary magnetic fields on ionic conduction processes in biological systems. Electromagnetic compatibility 1985. In: Proceedings of the 6th Symposium and Technical Exhibition on Electromagnetic Compatibility, Zurich, 5-7 March, pp. 205-210 (Paper 37G2).

TENFORDE, T.S., LEVY, L., & VERKLEROV, E. (1986a) Monitoring of circadian waveforms in rodents exposed to high-intensity static magnetic fields. I. Instrumentation. In: Proceedings of the 8th Annual Meeting of the Bioelectromagnetics Society, Madison, Wisconsin, 1-5 June, Gaithersburg, Maryland, Bioelectromagnetics Society.

TENFORDE, T.S., LEVY, L., & VEKLEROV, E. (1986b) Monitoring of circadian waveforms in rodents exposed to high-intensity static magnetic fields. II. Experimental results. In: Proceedings of the 8th Annual Meeting of the Bioelectromagnetics Society, Madison, Wisconsin, 1-5 June, Gaithersburg, Maryland, Bioelectromagnetics Society.

TEUCHER, I., BAESSLER, H., & LABES, M.M. (1971) Diffusion through neumatic liquid crystals. Nature phys. Sci., 229: 25.

THACH, J.S. (1968) A behavioural effect of intense DC electromagnetic fields. In: Vagtborg, H., ed. Use of non-human primates in drug evaluation, Austin, Texas, University of Texas Press, pp. 347-356.

THOMAS, A. & MORRIS, P.G. (1981) The effects of NMR exposure on living organisms. I. A microbial assay. Br. J. Radiol., 54: 615-621.

THOMPSON, S.P. (1909-10) A physiological effect of an alternating magnetic field. Proc. R. Soc. Lond. (Ser. B), 82: 396-398.

TOGAWA, T., OKAI, O., & OSHIMA, M. (1967) Observation of blood flow EMF in externally applied strong magnetic fields by surface electrodes. Med. biol. Eng., 5: 169-170.

TOMENIUS, L. (1986) 50-Hz electromagnetic environment and the incidence of tumours in Stockholm county. Bioelectromagnetics, 7: 191-207.

TOMENIUS, L., HELLSTROM, L., & ENANDER, B. (1982) Electrical constructions and 50 Hz magnetic field at the dwellings of tumour cases (0-18 years of age) in the county of Stockholm. In: Proceedings of the International Symposium and Occupational Health and Safety in Mining and Tunnelling, Prague, 21-25 June.

TORNQVIST, S., NORELL, S., AHLBOM, A., & KNAVE, B. (1986) Cancer in the electric power industry. Br. J. ind. Med., 43: 212-213.

TOROPTSEV, J.V. (1968) [Morphological changes and biological effects of magnetic field exposure.] Arkh. Patol., 30(3): 3-12 (in Russian).

TOROPTSEV, J.V. & SOLDATOVA, L.P. (1981) [Pathomorphological reactions of cerebrocortical neural elements to an alternating magnetic field.] Arkh. Patol., 43: 33-36 (in Russian with English summary).

TOROPTSEV, J.V., GARGANEYEV, G.P., GORSHENINA, T.I., & TEPLYAKOVA, N.L. (1974) Pathologoanatomic characteristics of changes in experimental animals under the influence of magnetic fields. In: Kholodov, YU.A., ed. Influence of magnetic fields on biological objects, Springfield, Virginia, National Technical Information Service, pp. 95-104 (NTIS Report No. HPRS 63038).

TUCKER, R.D. & SCHMITT, O.H. (1978) Tests for human perception of 60 Hz moderate strength magnetic fields. IEEE Trans. Biomed. Eng., 25: 509-518.

TVILDIANI, D.D., CHLAIDZE, T.I., DOLIDZE, N.V., GOLSAHVILI, L.N., & CHIKHLADZE, V.A. (1981) [Effects of experimental stationary magnetic fields on metabolism of some ions in blood and myocardium.] Soobsh. Akad. Nauk Gruz. SSR, 101: 169-172 (in Russian).

TVILDIANI, D.D., KURASHVILI, R.B., CHLAIDZYE, T.I., SELIKHOV, Y.V., & GAPRINDASHVILI, T.G. (1983) [Effect of constant electromagnetic fields on EKG parameters and homeostatis.] Soobshch. Akad. Nauk Gruz. SSR, 110: 413-416 (in Russian).

UBEDA, A., LEAL, J., TRILLO, M.A., JIMENEZ, M.A., & DELGADO, J.M.R. (1983) Pulse shape of magnetic fields influence chick embryogenesis. J. Anat., 137: 513-536.

UDINTSEV, N.A. & KHLYNIN, S.M. (1979) Effect of a variable magnetic field on activity of carbohydrate metabolism and tissue respiration in testicle tissue. Ukr. Biol. Zh., 50: 714-717.

UDINTSEV, N.A. & MOROZ, V.V. (1974) Response of the pituitary-adrenal system to the action of a variable magnetic field. Byull. Eksp. Biol. Med., 77: 51-52.

UDINTSEV, N.A. & MOROZ, V.V. (1982) [Function of hypophyseal-adrenal system under the effect of power frequency variable magnetic field of different regimes.] Gig. Tr. Prof. Zabol., 12: 54-56 (in Russian).

UDINTSEV, N.A., KANSKAIA, N.V., SHCHEPETIL'NIFOVA, A.I., ORDINA, O.M., & PICHURINA, R.A. (1976) [Dynamics of cardiac and skeletal muscle lactate dehydrogenase activity following a single exposure to an alternating magnetic field.] Byull.

Eksp. Biol. Med., 81: 670-672 (in Russian with English abstract).

UDINTSEV, N.A., SEREBROV, V.YU., & TSYROV, G.I. (1978) [Effects of industrial frequency variable magnetic fields upon thyroid gland function and thyroxin absorption in rats.] Byull. Eksp. Biol. Med., 11: 544-546 (in Russian).

UENO, S. & HARADA, K. (1982) Redistribution of dissolved oxygen concentration under strong DC magnetic fields. IEEE Trans. Mag., 18: 1704-1706.

UENO, S., MATSUMOTO, S., HARADA, K., & OOMURA, Y. (1978) Capacitive stimulatory effect in magnetic stimulation of nerve tissue. IEEE Trans. Mag., 14: 958-960.

UENO, S., LOVSUND, P., & OBERG, P.A. (1981) On the effect of alternating magnetic fields on action potential in lobster giant axon (Presented at 5th Nordic Meeting on Medical and Biological Engineering, Linkoping, Sweden, 10-13 June).

UENO, S., HARADA, K., & SHIOKAWA, K. (1984) The embryonic development of frogs under strong DC magnetic fields. IEEE Trans. Mag., 20: 1663-1665.

UENO, S., KITAHARA, T., HARADA, K., & SHIOKAWA, L. (1985) The effects of ELF magnetic and electric fields on the embryonic development of frogs (Presented at the 7th Annual Meeting of the Bioelectromagnetics Society, San Francisco, California, 16-20 June).

USSR (1970) Methodological recommendations. Occupational hygiene and prophylaxis of untoward action of magnetic fields on workers, Moscow, Ministry of Public Health.

USSR (1978) [Maximum permissible levels of exposure to static magnetic fields at work with magnetic installations and magnetic materials,] Moscow, Ministry of Public Health (Document No. 1742-77) (in Russian).

USSR (1985) Maximum permissible levels of magnetic fields with the frequency 50 Hz, Moscow, Ministry of Public Health (Document No. 3206-85).

VAGERO, D. & OLIN, R. (1983) Incidence of cancer in the electronics industry: using the new Swedish Environment Registry as a screening instrument. Br. J. ind. Med., 40: 188-192.

VAGERO, D., AHLBOM, A., OLIN, R., & SAHLSTEN, S. (1985) Cancer morbidity among workers in the telecommunications industry. Br. J. ind. Med., 42: 211-213.

VAINER, L.M., PODOPLELOV, A.V., LESHINA, T.V., SAGDEYEV, R.Z., & MOLIN, Y.N. (1978) [Effect of a magnetic field on the rate of decomposition of H_2O_2 by catalase and by the EDTA complex with Fe^{3+}.] Biofizika, 23: 234-242 (in Russian with English summary).

VAJDA, T. (1980) Investigation of magnetic field effect on trypsin activity. Radiat. environ. Biophys., 18: 275-280.

VALENTINUZZI, M. (1962) Theory of magnetophosphenes. Am. J. Med. Electr., 1: 112-121.

VALENTINUZZI, M. (1965) Notes on magnetic actions upon the nervous system. Bull. Math. Biophys., 27: 203-214.

VIKTOVA, L., FIALA, J., & PETZ, R. (1976) Effect of prolonged exposures to a magnetic field on the hematopoietic stem cell. Physiol. Bohemoslov., 25: 359-364.

VILENCHIK, M.M. (1982) [Magnetic susceptibility of rhodopsin.] Biofizika, 27: 31-36 (in Russian with English summary).

VYALOV, A.M. (1967) Magnetic fields as an environmental factor. Vestnik, 8: 52-58.

VYALOV, A.M. (1971) Clinico hygienic and experimental data on the effect of magnetic fields under occupational conditions, Moscow, Nauka, pp. 165-177.

VYALOV, A.M. (1974) Clinco-hygienic and experimental data on the effects of magnetic fields under industrial conditions. In: Kholodov, YU.A., ed. Influence of magnetic fields on biological objects, Springfield, Virginia, National Technical Information Service, pp. 163-174 (NTIS Report No. JPRS 63038).

VYALOV, A.M. & LISICHKINA, Z.S. (1966) [Definition of clinico-physiological shift in occupational personnel exposed to scattered static magnetic fields.] Gig. Tr. Prof. Zabol., 5: 39-43 (in Russian).

VYALOV, A.M., SHPIL'BERG, P.I., YUSHKEVICH, L.B., LISICHKINA, Z.S., RYABOVA, A.P., DMITRIYEVA, K.A., SOKOLOV, S.A., & ZVANILOVA, L.L. (1964) [To the question of static and variable magnetic field effects on human organism.] In:

[Scientific transactions,] Moscow, F.F. Erisman Hygiene Research Institute, pp. 169-175 (in Russian).

WACHTEL, H. (1979) Firing pattern changes and transmembrane current produced by extremely low frequency fields in pacemaker neurons. In: Phillips, R.D., Gillis, M.F., Kaune, W.T., & Mahlum, D.D., ed. Biological effects of extremely low frequency electromagnetic fields, Springfield, Virginia, US Department of Energy, National Technical Information Service, pp. 132-146 (NTIS Report No. CONF-781016).

WALCOTT, C., GOULD, J.L., & KIRSCHVINK, J.L. (1979) Pigeons have magnets. Science, 205: 1027-1029.

WALKER, M.M., KIRSCHVINK, J.L., CHANG, S.-B.R., & DIZON, A.E. (1984) A candidate magnetic sense organ in the Yellowfin tuna Thunnus albacares. Science, 224: 751-753.

WATSON, A.B., WRIGHT, J.S., & HIGHMAN, J. (1973) Electrical thresholds for ventricular fibrillation in man. Med. J. Aust., 1: 1179-1182.

WATSON, J. & DOWNES, E.M. (1978) The application of pulsed magnetic fields to the stimulation of bone healing in humans. Jpn. J. appl. Phys., 17: 215-217.

WELKER, H.A., SEMM, P., WILLIG, R.P., COMMENTZ, J.C., WILTSCHKO, W., & VOLLRATH, L. (1983) Effects of an artificial magnetic field on serotonin N-acetyltransferase activity and melatonin content of the rat pineal gland. Exp. Brain Res., 50: 426-432.

WERNER, H.-J., SCHULTEN, K., & WELLER, A. (1978) Electron transfer and spin exchange contributing to the magnetic field dependence of the primary photochemical reaction of bacterial photosynthesis. Biochim. Biophys. Acta, 502: 255-268.

WERTHEIMER, N. & LEEPER, E. (1979) Electrical wiring configurations and childhood cancer. Am. J. Epidemiol., 109(3): 273-284.

WERTHEIMER, N. & LEEPER, E. (1982) Adult cancer related to electrical wires near the home. Int. J. Epidemiol., 11(4): 345-355.

WHO (1981) EHC 16: Radiofrequency and microwaves, Geneva, World Health Organization.

WHO (1982) In: Suess, M.J., ed. Non-ionizing radiation protection, Copenhagen, World Health Organization, Regional Office for Europe (WHO Regional Publications, European Series No. 10).

WIKLUND, K., EINHORN, J., & EKLUND, G. (1981) An application of the Swedish cancer-environment registry. Leukaemia among telephone operators at the telecommunications administration in Sweden. Int. J. Epidemiol., 10: 373-376.

WIKSWO, J.P., Jr & BARACH, J.P. (1980) An estimate of the steady magnetic field strength required to influence nerve conduction. IEEE Trans. Biomed. Eng., 27: 722-723.

WILLIAMSON, S.J. & KAUFMAN, L. (1981) Biomagnetism. J. Mag. Mat., 22: 129-201.

WILTSCHKO, W. & WILTSCHKO, R. (1972) Magnetic compass of European robins. Science, 176: 62-64.

WINTERS, W.D., GUEST, G.F., WINTERS, B.T., & PHILLIPS, J.L. (1985a) Human leukocyte responses after exposure to 60 Hz electromagnetic fields in vitro (Presented at the 7th Annual Meeting of the Bioelectromagnetics Society, San Francisco, California, 16-20 June).

WINTERS, W.D., CRAWLEY, R., YOUNG, R.J., & PHILLIPS, J.L. (1985b) Biological responses of canine leukocytes after exposure to 60 Hz electromagnetic fields in vitro (Presented at the 7th Annual Meeting of the Bioelectromagnetics Society , San Francisco, California, 16-20 June).

WOLFF, S., CROOKS, L.E., BROWN, P., HOWARD, R., & PAINTER, R.B. (1980) Tests for DNA and chromosomal damage induced by nuclear magnetic resonance imaging. Radiology, 136: 707-710.

WORDSWORTH, O.J. (1974) Comparative long-term effects of liver damage in the rat after (a) localized x-irradiation and (b) localized x-irradiation in the presence of a strong homogeneous magnetic field. Radiat. Res., 57: 442-450.

WRIGHT, W.E., PETERS, J.M., & MACK, T.M. (1982) Leukaemia in workers exposed to electrical and magnetic fields. Lancet, 2(8308): 1160-1161.

YAROVITSKIY, M. (1986) Applications of magnets. Med. Gaz., 13(4562): 4.

YOUNG, R.W. (1978) Visual cells, daily rhythms, and vision research. Vision Res., 18: 573-578.

YOUNG, W. (1969) Magnetic field and in situ acetylcholinesterase in the vagal heart system. In: Barnothy, M.F., ed. Biological effects of magnetic fields, New York, London, Plenum Press, Vol. 2, pp. 79-102.

ZAFFANELLA, L.E. & DENO, D.W. (1978) Electrostatic and electromagnetic effects of ultra-high-voltage transmission lines, Palo Alto, California, Electric Power Research Institute (Final report EPRI EL-802).

ZOEGER, J., DUNN, J.R., & FULLER, M. (1981) Magnetic material and the head of the common Pacific dolphin. Science, 213: 892-894.